# Lecture Notes in Computer Science 14401

The series Lecture Notes in Computer Science (LNCS), including its subseries Lecture Notes in Artificial Intelligence (LNAI) and Lecture Notes in Bioinformatics (LNBI), has established itself as a medium for the publication of new developments in computer science and information technology research, teaching, and education.

LNCS enjoys close cooperation with the computer science R & D community, the series counts many renowned academics among its volume editors and paper authors, and collaborates with prestigious societies. Its mission is to serve this international community by providing an invaluable service, mainly focused on the publication of conference and workshop proceedings and postproceedings. LNCS commenced publication in 1973.

Nina Gierasimczuk ·
Fernando R. Velázquez-Quesada
Editors

# Dynamic Logic

## New Trends and Applications

5th International Workshop, DaLí 2023
Tbilisi, Georgia, September 15–16, 2023
Revised Selected Papers

 Springer

*Editors*
Nina Gierasimczuk ⓘD
Technical University of Denmark
Lyngby, Denmark

Fernando R. Velázquez-Quesada ⓘD
University of Bergen
Bergen, Norway

ISSN 0302-9743          ISSN 1611-3349 (electronic)
Lecture Notes in Computer Science
ISBN 978-3-031-51776-1          ISBN 978-3-031-51777-8 (eBook)
https://doi.org/10.1007/978-3-031-51777-8

This Springer imprint is published by the registered company Springer Nature Switzerland AG
The registered company address is: Gewerbestrasse 11, 6330 Cham, Switzerland

Paper in this product is recyclable.

# Preface

Building on the ideas of Floyd-Hoare logic, Dynamic Logic (*DL*) was introduced in the 1970s as a formal tool for reasoning about and verifying classic imperative programs. Its main feature is that it allows us to reason not only about the states in which a program might be, but also about the structured actions that define transitions from one state to another. Over time, *DL*'s aim has evolved and expanded. It can be seen now as a general set of ideas, tools and techniques devised for representing, describing and reasoning about diverse kinds of actions, including frameworks tailored for specific programming problems and paradigms (e.g., separation logics), settings for modelling new computing domains (e.g., probabilistic, continuous and quantum computation), frameworks for reasoning about information dynamics (e.g., dynamic epistemic logics) and systems for reasoning about long-term information dynamics (e.g., learning theory).

Both its theoretical relevance and practical potential make *DL*s a topic of interest in various scientific venues, from wide-scope software engineering conferences to events specific to modal logics. The aim of the *Dynamic Logic – New Trends and Applications* (DaLí) Workshop is to bring together researchers who share an interest in the formal study of actions, to create a platform for presenting ongoing research, to foster discussions and encourage new collaborations. Previous editions of DaLí took place in Brasília (2017), Porto (2019) and online (2020, 2022). In 2023, DaLí (https://dali2023.compute.dtu.dk/)[1] took place on the 15th and 16th of September in Tbilisi, Georgia, at Ivane Javakhishvili Tbilisi State University.

Apart from contributed presentations, the workshop featured two invited lectures. In her presentation *Standard and General Completeness of Modal Many-valued Logics*, Amanda Vidal compared modal logics evaluated over standard algebras and those evaluated over arbitrary ones of the corresponding variety. The main results concerned the local and global logical entailments of the modal Gödel, Łukasiewicz and product logics. Igor Sedlár's invited lecture was titled *Kleene Algebras for Weighted Programs*. He presented different semantics for (the propositional abstraction of) weighted programs (a generalisation of probabilistic programs): a model based on weighted sets of guarded strings, a relational model based on weighted relations in a state space, and, finally, a more general setting of functional semantics based on functions from multi-monoids to quantales.

The call for papers for the DaLí Workshop was answered by a significant number of contributions, and each submission received three double-blind reviews. After careful consideration, nine submissions were accepted for presentation. This volume contains revised versions of eight of the contributed papers, whose descriptions can be found below.

In the paper *A Spatial Logic with Time and Quantifiers*, Laura Bussi, Vincenzo Ciancia and Fabio Gadducci extend a spatial logic for closure spaces in two ways: by adding quantification on both individuals and atomic propositions and by adding

---

[1] Website kindly hosted by DTU Compute, Denmark.

temporal operators. The resulting setting is a quantified spatio-temporal logic that can be used to state spatial properties, possibly involving the identity of individuals, in models that evolve over time.

In *Logic of the Hide and Seek Game: Characterization, Axiomatization, Decidability*, Qian Chen and Dazhu Li provide new results about LHS, a logic that describes the children's game of hide and seek, and whose satisfiability problem has been shown to be undecidable when the setting contains multiple relations and the language contains an equality constant. First, they provide a characterization theorem for the expressive power of the logic. Second, they show that the satisfiability problem remains undecidable even with a single relation. Finally, they study LHS⁻, a fragment without the equality constant, providing an axiom system and showing that its satisfiability problem is decidable.

The paper *Axiomatization of Hybrid Logic of Link Variations* by Penghao Du and Qian Chen investigates a family of dynamic logics whose 'link variations' modalities describe the effects of deleting, adding or swapping edges in a relational structure. Thanks to the use of hybrid logic operators, the authors provide complete axiomatizations for both global and local versions of the modalities.

In *Kleene Algebra of Weighted Programs With Domain*, Igor Sedlár extends the work on the propositional abstraction of weighted programs presented in his invited talk. The setting now includes a weighted weakest precondition, which is formalized using a weak version of the domain operator of Kleene algebra with domain.

*Automated Quantum Program Verification in Dynamic Quantum Logic* introduces an automated approach for the verification of quantum programs. The authors, Tsubasa Takagi, Canh Minh Do and Kazuhiro Ogata, formalise notions such as quantum states, quantum gates and projections, and use laws from quantum mechanics and matrix operations to reason about quantum computation. Formal verification is then conducted through an equational simplification process.

*Predictive Theory of Mind Models Based on Public Announcement Logic* continues research on the use of epistemic logic for studying Theory of Mind, the cognitive capacity to attribute internal mental states (e.g., knowledge, beliefs, intentions) to oneself and others. In this work, Jakob Dirk Top, Catholijn Jonker, Rineke Verbrugge and Harmen de Weerd use public announcement logic to propose a logic of bounded Theory of Mind. Then, they use this logic and statistical methods to estimate the distribution of 'Theory of Mind levels' in a population of human participants of a behavioural experiment.

In *Learning by Intervention in Simple Causal Domains*, Katrine Bjørn Pedersen Thoft and Nina Gierasimczuk propose a framework for learning causal dependencies between variables in an environment with causal relations. The setting is a simplification of the standard causal framework in which the environment is fully observable and the underlying causal structure is of a simple nature. The authors propose two learning methods that make use of causal frames, causal models and dependence models: a learning function to handle finite identifiability of single-variable simple dependence, and a learning algorithm to handle a more general multi-variable notion of dependence.

In *A Logical Approach to Doxastic Causal Reasoning*, Kaibo Xie, Qingyu He and Fenrong Liu propose a formal logic that brings together belief revision and causal reasoning. Their proposal introduces a causal plausibility model which combines ideas from plausibility models for representing beliefs with ideas from structural equation models for representing causal knowledge. Besides presenting examples that show the use of the framework, the authors present a complete axiomatization, as well as a decidability result.

We present this volume in the hope that the research community will find the contents interesting and inspiring. We are very glad that DaLí 2023 turned out very interdisciplinary and featured a broad range of topics. We are very grateful not only to the authors who submitted their original work, but also to all members of the Program and Organizing Committees, as well as the invited speakers and all the participants. This publication would not have been possible without the commitment of their time and work.

November 2023                                    Nina Gierasimczuk
                                     Fernando R. Velázquez-Quesada

# Organization

## Program Committee Chairs

Nina Gierasimczuk           Technical University of Denmark, Denmark
Fernando R. Velázquez-Quesada    Universitetet i Bergen, Norway

## Steering Committee

Alexandru Baltag        University of Amsterdam, The Netherlands
Luís Barbosa        University of Minho, Portugal
Mário Benevides        Federal University of Rio de Janeiro, Brazil
Johan van Benthem        University of Amsterdam, The Netherlands
Hans van Ditmarsch        CNRS, IRIT and Toulouse University, France
Nina Gierasimczuk        Technical University of Denmark, Denmark
David Harel        Weizmann Institute of Science, Israel
Andreas Herzig        IRIT and University Paul Sabatier, France
Dexter Kozen        Cornell University, USA
Fenrong Liu        Tsinghua University, China
Alexandre Madeira        University of Aveiro, Portugal
Igor Sedlár        Czech Academy of Sciences, Czech Republic
Manuel Martins        University of Aveiro, Portugal
Sonja Smets        University of Amsterdam, The Netherlands
Vaughan Pratt        Stanford University, USA

## Program Committee

Natasha Alechina        Utrecht University, The Netherlands
Carlos Areces        Universidad Nacional de Córdoba, Argentina
Phillippe Balbiani        IRIT and Toulouse University, France
Alexandru Baltag        University of Amsterdam, The Netherlands
Luís Barbosa        University of Minho, Portugal
Mário Benevides        Federal University of Rio de Janeiro, Brazil
Johan van Benthem        University of Amsterdam, The Netherlands
Thomas Bolander        Technical University of Denmark, Denmark
Zoé Christoff        University of Groningen, The Netherlands
Stéphane Demri        CNRS, France

| Hans van Ditmarsch | CNRS, IRIT and Toulouse University, France |
| Raul Fervari | Universidad Nacional de Córdoba, Argentina |
| Sabine Frittella | LIFO, France |
| Asta Halkjær From | Technical University of Denmark, Denmark |
| Malvin Gattinger | University of Amsterdam, The Netherlands |
| Sujata Gosh | ISI Chennai, India |
| Davide Grossi | University of Groningen, The Netherlands |
| Andreas Herzig | IRIT and University Paul Sabatier, France |
| Thomas Icard | Stanford University, USA |
| Gabriele Kern-Isberner | Technical University of Dortmund, Germany |
| Sophia Knight | University of Minnesota, USA |
| Alexandre Madeira | University of Aveiro, Portugal |
| Eric Pacuit | University of Maryland, USA |
| Alessandra Palmigiano | Vrije University of Amsterdam, The Netherlands |
| Elaine Pimentel | University College London, UK |
| Carlo Proietti | ILC, Italy |
| Vít Punčochář | Czech Academy of Sciences, Czech Republic |
| Rasmus Rendsvig | University of Copenhagen, Denmark |
| Mehrnoosh Sadrzadeh | University College London, UK |
| Katsuhiko Sano | Hokkaido University, Japan |
| Igor Sedlár | Czech Academy of Sciences, Czech Republic |
| Sonja Smets | University of Amsterdam, The Netherlands |
| Sheila Veloso | Federal University of Rio de Janeiro, Brazil |
| Fan Yang | Utrecht University, The Netherlands |
| Thomas Ågotnes | Universitetet i Bergen, Norway |

## Local Organiser

| David Gabelaia | A. Razmadze Mathematical Institute, Georgia |

## Additional Reviewers

Avijeet Ghosh
Tim French
Mattia Panettiere
Wojciech Rozowski

# Contents

# A Spatial Logic with Time and Quantifiers

Laura Bussi[1,2]($\boxtimes$), Vincenzo Ciancia[2], and Fabio Gadducci[1]

[1] Department of Computer Science, University of Pisa, Pisa, Italy
laura.bussi@phd.unipi.it, fabio.gadducci@unipi.it
[2] CNR-ISTI, Pisa, Italy
{l.bussi,v.ciancia}@isti.cnr.it

**Abstract.** Spatial logics are formalisms for expressing topological properties of structures based on geometrical entities and relations. In this paper we consider SLCS, the Spatial Logic for Closure Spaces, recently used for describing features of images and video frames. We extend SLCS in two directions. We first introduce first-order quantifiers, ranging on both individuals and atomic propositions. We then equip the logic with temporal operators, and provide a linear-time semantics over finite traces. The resulting formalism allows to state properties about geometrical entities whose attributes change along time. For both extensions, we prove the equivalence of their operational semantics with a denotational one.

## 1 Introduction

Spatial logics are formalisms for expressing topological properties of structures based on geometrical entities and relations, and as such have been extensively studied since the first half of the last century [1]. Recently, such logics have been further explored for the modelling of computational devices, ranging from collective adaptive [13,14] and cyber-physical systems [22,24] to pattern synthesis [5].

Introduced in [16], the Spatial Logic for Closure Spaces (SLCS) uses as models a generalisation of topological spaces, known as *pretopological* or *Čech closure spaces*. These spaces include interesting structures such as binary relations/simple graphs. And since images can be interpreted as graphs, whose structure is given by pixels with a chosen adjacency relation, the SLCS model checker VoxLogicA [8] has been used for the analysis of 2D/3D pictures, in particular for the problem of "contouring" in medical imaging [4,7].

Supported by University of Pisa project PRA_2022_99 "FM4HD", MUR project PRIN 20228KXFN2 "STENDHAL", CNR (Italy) and SRNSFG (Georgia) bilateral project CNR-22-010 "Model Checking for Polyhedral Logic", and European Union - Next Generation EU - MUR project PNRR PRI ECS00000017 PRR.AP008.003 "THE - Tuscany Health Ecosystem". The authors thank Diego Latella and Mieke Massink for fruitful discussions on spatio-temporal logics and their applications.

N. Gierasimczuk and F. R. Velázquez-Quesada (Eds.): DaLí 2023, LNCS 14401, pp. 1–19, 2024.
https://doi.org/10.1007/978-3-031-51777-8_1

SLCS has proved to be quite expressive in characterising the structural properties of a graph. However, it does not possess operators for constructing *named references* to "individuals"—be these points, regions, atomic propositions, or agents moving in space. For instance, one might ask if there is a region $X$ of an image, satisfying a given logical property, which in some time will become larger than another one. This kind of analysis has immediate applications in medical imaging for *lesion tracking*, focussing on the temporal evolution of a lesion in a series of snapshots of a patient's situation (a "longitudinal study"). In this work, we develop the ideas of [16] and [11], adopting the same setting of [17] to model spatio-temporal situations. First of all, we provide a precise correspondence between spaces and relations, streamlining various results discussed in the literature on SLCS. We also present a succinct syntax of SLCS, including just the backward $\overleftarrow{\rho}$ and forward $\overrightarrow{\rho}$ reachability operators, which reflect the well-known *until* operator of temporal logic and have efficient model checking algorithms in VoxLogicA. Such operators allow to state properties of points of space akin to *there is a finite path from point $x_1$ to some point $x_2$, such that $x_2$ satisfies a given formula $\phi_2$, and the path passes only through points satisfying another formula $\phi_1$.* Taking inspiration from [11], we introduce two extensions of SLCS. The first one concerns first-order quantification, which may predicate on points of a space and the atomic propositions they may satisfy. The second introduces temporal operators, similar in spirit to [14]. Finally, these extensions are merged, distilling an expressive and flexible quantified spatio-temporal logic.

*A Running Example: Video Stream Analysis.* The logic we propose allows to state properties involving the identity of a node, in a graph whose structure does not change, yet the propositions holding at each node may. Throughout the paper, we illustrate its expressiveness by a simple example: the analysis of video streams, demonstrated using the well-known Pac-Man™ videogame. The example is taken from [12], where only purely spatial properties were considered.

Pac-Man is a 2D video game released by the Japanese firm Bandai-Namco in 1980. It has a simple, yet interesting structure: the main character of the game, Pac-Man, moves inside a maze. Along the corridors, several peach dots are placed, together with four energiser pellets positioned in the corners. Furthermore, four coloured ghosts (Inky, Blinky, Pinky, and Clyde) try to capture Pac-Man, moving in the maze according to different routines. A twist happens when Pac-Man eats an energiser pellet: in this case, the ghosts' colours turn to blue, and they can be caught by Pac-Man instead. The aim of a single level is to eat all the dots and pellets, avoiding to be captured by a ghost.

Despite its simplicity, the Pac-Man videogame is a clear example of applicability of our logical framework. The spatial structure does not change along time: the graph underlying each video frame is always the same. Instead, atomic properties associated to a node/pixel, that is, the colours, vary along time: for example, Pac-Man is represented by yellow-coloured pixels that are inside the maze (note that there are other areas with the same colour, representing the remaining lives, see Fig. 1). Such a setting is useful in real-world applications. Consider, for instance, lesion tracking in medical imaging. The input data are *snapshots* of a patient at different times. After what is called the *co-registration* phase, all images have the same structure (resolution and physical dimensions). In other words, the underlying graph never changes, while the colours of the pixels, i.e. the atomic propositions, change along the temporal axis.

*Related Work.* The task to investigate quantification in modal logic interpreted over spaces was already tackled in various works. An important example are the works by Awodey and Kishida [2,20], where first order modal logic is provided with a topological interpretation. The proposed approach is quite different from ours: in this case, sheaves are used to combine denotational semantics of modal logic and first order logic, and quantification is permitted only over points. Moreover, this approach applies only to topological spaces.

Spatio-temporal reasoning has also been a topic of interest along years, and various approaches have been proposed to combine space and time. Products of modal logics have been considered to this end [9]. Products of modal logics give rise to *multi-modal* logic languages, where different modal operators can be used to reason about different aspects of a model (in this case, the spatial and temporal aspects). Despite the fact that we also consider products of modal logics, the cited proposal is quite different. Again in this case, only topological structures are considered, and the temporal fragment is interpreted over the pair $(\mathbb{N}, <)$, thus being equivalent to the classic PTL temporal logic. In our case, instead, we only consider interpretation over finite traces. A comprehensive study of spatio-temporal approaches to modal logics is given by [17], where various kinds of spaces (e.g. Euclidean or Aleksandroff) are considered. This work offers an interesting study of the tradeoff between expressivity and complexity of various spatio-temporal logic, and it is our main reference for state-of-the-art languages that combine space and time. Still, the topic of the considered logics is topological spaces, thus lacking the generality that we aim to have.

Closer to our proposal, and in some sense orthogonal to it, is the one developed in [14], where branching time operator where introduced and no quantification was considered. In this case, the language was developed to reason about evolving smart systems (e.g. bike sharing systems), thus a branching time logic was adopted for the temporal part. We drop this kind of approach in favour of linear time operators, which are more likely to be useful in a setting of medical imaging, where we state properties about a set of images on a single timeline.

*Synopsis.* The structure of the paper follows. Section 2 gives an overview of the models currently used for SLCS and we recast them uniformly, making precise the correspondence with binary relations/simple graphs. Section 3, presents a succinct version of SLCS, which is equipped with existential quantifiers in Sect. 4 and with linear-time operators in Sect. 5. Finally, Sect. 6 proposes a quantified spatio-temporal logic. Each section gives the correspondence between the semantics with respect to a single spatial path/temporal trace and a denotational one, and it is rounded up with an instance of our running example. Section 7 closes the paper, summing up our results and hinting at future works.

## 2   Some Notions on Spaces and Relations

We recall some notions related to spaces, used as domains of interpretation of various logics (see [1]) including SLCS, and discuss their links with binary relations/simple graphs, making precise remarks scattered in papers on SLCS.

### 2.1   Preliminaries on Spaces

We open by listing some basic properties and definitions for spaces.

**Definition 1.** *A* space $\mathcal{C}$ *is a pair* $(S, C)$ *such that* $S$ *is a set of points and* $C : 2^S \to 2^S$ *is a function satisfying* $C(\emptyset) = \emptyset$ *and* $C(X \cup Y) = C(X) \cup C(Y)$ *for* $X, Y \subseteq S$. *A space is* complete *if* $C(\bigcup_{i \in I} X_i) = \bigcup_{i \in I} C(X_i)$ *for any* $I$.

If $S$ is finite then a space $(S, C)$ is always complete. Given a space $(S, C)$ and a subset $X \subseteq S$, we denote the complement $S \backslash X$ of $X$ in $S$ as $X^c$. And while $C$ is called the closure operator, its dual is the interior $I(X) = C(X^c)^c = S \backslash C(S \backslash X)$.

**Definition 2.** *A* space $(S, C)$ *is* pre-topological *if* $X \subseteq C(X)$ *holds for all* $X \subseteq S$; *it is* Alexandrov *if it is pre-topological and complete; and it is* topological *if it is pre-topological and* $C(C(X)) \subseteq C(X)$ *holds for all* $X \subseteq S$.

The notions above are standard from the literature on topology. In the literature on spatial logics, pre-topological and Alexandrov spaces are called Cêch closure spaces and quasi-discrete Cêch closure spaces, respectively.

Note that for any space we can define a sort of inverse $\mathcal{C}^{-1} = (S, C^{-1})$, for $C^{-1}(X) = \bigcup_{x \in X} \{y \mid x \in C(\{y\})\}$, which is complete by definition. In order to identify those cases where a space and its inverse interact properly, we take inspiration from modal algebras and introduce the notion of conjugate spaces.

**Definition 3.** *Two spaces* $(S, C_1)$ *and* $(S, C_2)$ *are* conjugate *if they satisfy* $X \subseteq I_1(C_2(X)) \cap I_2(C_1(X))$.

*Remark 1.* The law for conjugate spaces can be stated as "$C_1(X) \subseteq Y$ iff $X \subseteq C_2(Y)$", which explicitly tells that the two closures are the respective inverses.

**Proposition 1.** *Let* $\mathcal{C}$ *be a complete space. Then* $\mathcal{C}$ *and* $\mathcal{C}^{-1}$ *are conjugate.*

*Proof.* We just need to prove that for any $X, Y$ we have that $C(X) \cap Y = \emptyset$ iff $X \cap C^{-1}(Y) = \emptyset$. Now, let us assume that $C(X) \cap Y = \emptyset$ and there exists $x$ such that $x \in X \cap C^{-1}(Y)$. Thus $x \in X$ and $x \in C^{-1}(Y)$. By definition, $x \in C^{-1}(Y)$ implies that there exists $y \in Y$ such that $x \in C^{-1}(\{y\})$, that is, $y \in C(\{x\})$, hence $y \in C(X)$ since $C$ is complete, thus $y \in C(X) \cap Y$, a contradiction. The inverse direction is analogous.

*Remark 2.* Note that we cannot drop the completeness requirement for $\mathcal{C}$ in the proposition above. Consider e.g. the set $\mathbb{N}$ of natural numbers and a function $C : 2^{\mathbb{N}} \to 2^{\mathbb{N}}$ such that $C(X) = \emptyset$ if $X$ is either empty or finite, and $C(X) = \mathbb{N}$ if $X$ is infinite. Clearly, $(\mathbb{N}, C)$ is a space, albeit not complete. Now, we have that $C(\{n\}) = \emptyset$ for all $n \in \mathbb{N}$, so that $C^{-1}(\{m\}) = \{n \mid m \in C(\{n\})\} = \emptyset$ for all $m \in \mathbb{N}$, which implies that $C^{-1}(Y) = \emptyset$ for all $Y \subseteq \mathbb{N}$. Thus, for any infinite set $X \subseteq \mathbb{N}$, we have that $C(X) \cap Y = Y$ while $X \cap C^{-1}(Y) = \emptyset$.

## 2.2   Spaces vs. Relations

There is a reason to focus on complete spaces, namely, the fact that they have a tight connection with binary relations (i.e. simple graphs/unlabelled Kripke frames). In the following we consider relations on a set $S$: we identify them as functions $R : S \to 2^S$ and denote $2^R : 2^S \to 2^S$ the lifting $2^R(X) = \bigcup_{x \in X} R(x)$.

Now, each space $\mathcal{C} = (S, C)$ induces a relation $R_\mathcal{C} : S \to 2^S$ defined as $R_\mathcal{C}(x) = C(\{x\})$. Note that for any finite $X \subseteq S$ it holds $2^{R_\mathcal{C}}(X) = C(X)$, and the equality holds also for infinite $X$ if $\mathcal{C}$ is complete. Vice versa, each relation $R : S \to 2^S$ induces a complete space $\mathcal{C}_R = (S, C_R)$ defined as $C_R(X) = 2^R(X)$.

**Lemma 1.** *Let $R : S \to 2^S$ be a relation. Then $R_{\mathcal{C}_R}(x) = R(x)$ for all $x \in S$. Let $\mathcal{C}$ be a complete space. Then $C_{R_\mathcal{C}}(X) = C(X)$ for all $X \subseteq S$.*

Thus, interpreting logics on complete spaces is the same as using as models the underlying relations. What is also noteworthy is that some laws holding for complete spaces turn out to state structural properties of such relations.

**Proposition 2.** *Let $\mathcal{C}$ be a complete space and $R_\mathcal{C}$ the associated relation. Then*

- *$\mathcal{C}$ satisfies $X \subseteq C(X)$ iff $R_\mathcal{C}$ is reflexive*
- *$\mathcal{C}$ satisfies $C(C(X)) \subseteq C(X)$ iff $R_\mathcal{C}$ is transitive*
- *$\mathcal{C}$ satisfies $X \subseteq I(C(X))$ iff $R_\mathcal{C}$ is symmetric*

*Proof.* The first two items are kind of obvious thanks to Proposition 1. Thus, let us now look at the third property. For $R_\mathcal{C}$ being symmetric means that for all $x, y$ it holds that $y \in R_\mathcal{C}(x)$ iff $x \in R_\mathcal{C}(y)$ or, equivalently, that $y \notin R_\mathcal{C}(x)$ iff $x \notin R_\mathcal{C}(y)$. Satisfying $X \subseteq I(C(X))$ means that $X \subseteq C(C(X)^c)^c$. Recall now that for a complete space we have $2^{R_\mathcal{C}}(X) = C(X)$, and for the sake of calculations consider the relation $D(x) = S \setminus R_\mathcal{C}(x)$. Thus, axiom $X \subseteq I(C(X))$ can be expressed as $X \subseteq C(\bigcap_{x \in X} D(x))^c = \bigcap_{z \in \bigcap_{x \in X} D(x)} D(z)$.

($\Longrightarrow$) Let us assume that there exist $x, y$ such that $x \in R_\mathcal{C}(y)$ and $y \in D(x)$. Assuming $X = \{x\}$, the axiom becomes $x \in \bigcap_{z \in D(x)} D(z)$. Since $y \in D(x)$, the axiom implies $x \in D(y)$, which contradicts $x \in R_\mathcal{C}(y)$.

($\Longleftarrow$) Let us assume that $R_\mathcal{C}$ is symmetric and that there exists $X$ such that $X \not\subseteq I(C(X))$. The latter means that there exists $y \in X$ such that $y \notin I(C(X))$. So, there exists $w \in \bigcap_{x \in X} D(x)$ such that $y \notin D(w)$, i.e. $y \in R_\mathcal{C}(w)$. By symmetry $w \in R_\mathcal{C}(y)$, that is, $w \notin D(y)$, which contradicts $w \in \bigcap_{x \in X} D(x)$.

Finally, recall how for a space $(S, C)$ we defined a kind of inverse space $(S, C^{-1})$, inspired by the analogous notion for relations: in fact, given $R : S \to 2^S$, its inverse $R^{-1} : S \to 2^S$ is the relation such that $R^{-1}(x) = \{y \mid x \in R(y)\}$.

**Proposition 3.** *Let $(S, C)$ be a space. Then $R_C^{-1} = R_{C^{-1}}$.*

## 3   Spatial Logics

This section recalls syntax and semantics of spatial logics (SL), introduces its denotational semantics, and makes precise its connection with CTL.

We start by assuming a set $P$ of atomic propositions, ranged over by $a, b, \ldots$

**Definition 4.** *The formulae $\Phi$ of SL are given by the grammar*

$$\Phi ::= \texttt{true} \mid a \mid \neg\Phi \mid \Phi \wedge \Phi \mid \vec{\rho}\, \Phi[\Phi] \mid \overleftarrow{\rho}\, \Phi[\Phi]$$

We denote the Boolean operators $\texttt{false} = \neg\texttt{true}$ and $(\Phi \vee \Phi) = \neg(\neg\Phi \wedge \neg\Phi)$. We also denote $\vec{\mathcal{N}}\, \Phi = \vec{\rho}\, \Phi[\texttt{false}]$ and $\overleftarrow{\mathcal{N}}\, \Phi = \overleftarrow{\rho}\, \Phi[\texttt{false}]$, which for our models are the equivalent of next and previous in temporal logics (as made precise later).

Let us now consider the semantics. Since we focus on complete spaces, we may equivalently describe our models in terms of relations. Thus, a model $\mathcal{T}$ is a four-tuple $\langle S, R, P, L \rangle$ such that $S$ is a set of points, $R : S \rightarrow 2^S$ a relation, $P$ a set of atomic propositions, and $L : P \rightarrow 2^S$ a labelling function. We also define the standard notion of spatial path in $\mathcal{T}$ from point $s_0$ to point $s_n$, i.e., a sequence $s_0 \ldots s_n$ with $n \geq 1$ such that $s_i \in R(s_{i-1})$ for all $i = 1 \ldots n$.

**Definition 5.** *Let $\mathcal{T}$ be a model. The semantics of a SL formula $\Phi$ with respect to a point $s \in S$ is given by the rules*

- $s \models \texttt{true}$
- $s \models a$ *if* $s \in L(a)$
- $s \models \neg\Phi$ *if* $s \not\models \Phi$
- $s \models \Phi_1 \wedge \Phi_2$ *if* $s \models \Phi_1$ *and* $s \models \Phi_2$
- $s \models \vec{\rho}\, \Phi_1[\Phi_2]$ *if there exists a spatial path* $s s_1 \ldots s_n$ *in* $\mathcal{T}$ *such that* $s_n \models \Phi_1$ *and* $s_j \models \Phi_2$ *for all* $j = 1 \ldots n - 1$
- $s \models \overleftarrow{\rho}\, \Phi_1[\Phi_2]$ *if there exists a spatial path* $s_0 \ldots s_{n-1} s$ *in* $\mathcal{T}$ *such that* $s_0 \models \Phi_1$ *and* $s_j \models \Phi_2$ *for all* $j = 1 \ldots n - 1$

The derived Boolean operators behave as expected, e.g. $s \not\models \texttt{false}$ for all states $s$. We recover the intuitive meaning of $\vec{\mathcal{N}}\, \Phi$ (hence, the existence of a direct connection between two points) as $\vec{\rho}\, \Phi[\texttt{false}]$, since $s \models \vec{\rho}\, \Phi[\texttt{false}]$ is equivalent to say that $s_1 \models \Phi$ for some $s_1 \in R(s)$. Similarly for $\overleftarrow{\mathcal{N}}$ with respect to $R^{-1}$. Finally, note that $\vec{\mathcal{N}}$ and $\overleftarrow{\mathcal{N}}$ distribute over the Boolean disjunction operator, so that e.g. $s \models \vec{\mathcal{N}}(\Phi_1 \vee \Phi_2)$ iff $s \models (\vec{\mathcal{N}}\, \Phi_1) \vee (\vec{\mathcal{N}}\, \Phi_2)$.

**Lemma 2.** *Let $\mathcal{T}$ be a model, $s \in S$ a point, and $\Phi_1, \Phi_2$ SL formulae. Then $s \models \vec{\rho}\, \Phi_1[\Phi_2]$ iff $s \models \vec{\mathcal{N}}\, \Phi_1 \vee \vec{\mathcal{N}}(\Phi_2 \wedge \vec{\rho}\, \Phi_1[\Phi_2])$ (and similarly for $\overleftarrow{\rho}\, \Phi_1[\Phi_2]$).*

*Proof.* Let $s \models \vec{\rho}\, \Phi_1[\Phi_2]$. It holds if there exists a path $s s_1 \ldots s_n$ in $\mathcal{T}$ such that $s_n \models \Phi_1$ and $s_j \models \Phi_2$ for all $j = 1 \ldots n - 1$. Let us assume that $n = 1$. This is equivalent to say that $s_1 \models \Phi_1$, hence $s \models \vec{\mathcal{N}}\, \Phi_1$. So, let $n > 1$. This means that $s_1 \models \Phi_2$, $s_n \models \Phi_1$, and $s_j \models \Phi_2$ for all $j = 2 \ldots n - 1$, which is in turn equivalent to state that $s \models \vec{\mathcal{N}}(\Phi_2 \wedge \vec{\rho}\, \Phi_1[\Phi_2])$.

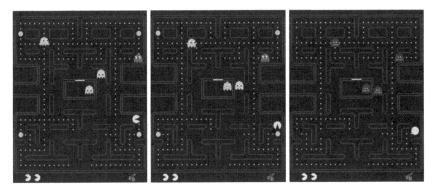

**Fig. 1.** A sequence of Pac-Man frames: ghosts turn to blue immediately after frame 2. (Color figure online)

*Example 1.* Consider our running example, in particular the first frame of Fig. 1. As said above, we assume we have a set of atomic propositions $AP$ denoting colours. There is only one area satisfying the formula *orange*, namely the orange ghost. On the other hand, three different areas satisfy *yellow* and, for the moment being, we are not able to distinguish the active Pac-Man from the ones representing the remaining lives. However, we can already check an interesting property. So, let $ghost = orange \lor pink \lor lightBlue \lor red$. The pixels of a Pac-Man that is going to be caught by a ghost are identified via the formula $yellow \land \vec{\rho} ghost[yellow]$. Such formula finds all the yellow pixels that are connected, via a path of yellow ones (except the last one, see Definition 5), to a pixel belonging to a ghost. Indeed, no such pixel exists in the three frames considered.

### 3.1 Denotational Semantics of SL

The denotational meaning of a formula $\Phi$ is going to be a set of points in our model $\mathcal{T}$. The interpretation of the Boolean and the next and previous step operators is immediate: only the reachability operators need some care.

**Definition 6.** *Let $\mathcal{T}$ be a model. The denotational semantics of a SL formula $\Phi$ is given by the rules*

- $[\![\mathtt{true}]\!] = S$
- $[\![a]\!] = L(a)$
- $[\![\neg\Phi]\!] = [\![\Phi]\!]^c = S \setminus [\![\Phi]\!]$
- $[\![\Phi_1 \land \Phi_2]\!] = [\![\Phi_1]\!] \cap [\![\Phi_2]\!]$
- $[\![\vec{\mathcal{N}}\,\Phi]\!] = 2^{R^{-1}}([\![\Phi]\!]) = \{s \in S \mid R(s) \cap [\![\Phi]\!] \neq \emptyset\}$
- $[\![\overleftarrow{\mathcal{N}}\,\Phi]\!] = 2^{R}([\![\Phi]\!]) = \{s \in S \mid R^{-1}(s) \cap [\![\Phi]\!] \neq \emptyset\}$
- $[\![\vec{\rho}\,\Phi_1[\Phi_2]]\!] = \mathtt{lfp}_Z\,([\![\vec{\mathcal{N}}\,\Phi_1]\!] \cup [\![\vec{\mathcal{N}}\,(\Phi_2 \land Z)]\!])$
- $[\![\overleftarrow{\rho}\,\Phi_1[\Phi_2]]\!] = \mathtt{lfp}_Z\,([\![\overleftarrow{\mathcal{N}}\,\Phi_1]\!] \cup [\![\overleftarrow{\mathcal{N}}\,(\Phi_2 \land Z)]\!])$

The semantics associates a set of points to a formula. The interpretation of the $\vec{\mathcal{N}}$ and $\bar{\mathcal{N}}$ operators is clearly monotone with respect to subset inclusion, thus the least fix-point in the semantics of the $\vec{\rho}$ and $\bar{\rho}$ operators are well-defined.

*Remark 3.* For the sake of simplicity, in Definition 6 we considered $\vec{\mathcal{N}}$ and $\bar{\mathcal{N}}$ as primitive operators, instead of derived ones. However, it is easy to see that $[\![\vec{\rho}\,\Phi[\mathtt{false}]]\!] = \mathtt{lfp}_Z([\![\vec{\mathcal{N}}\,\Phi]\!] \cup ([\![\vec{\mathcal{N}}\,(\mathtt{false} \wedge Z)]\!])) = [\![\vec{\mathcal{N}}\,\Phi]\!]$, and analogously $[\![\bar{\rho}\,\Phi[\mathtt{false}]]\!] = [\![\bar{\mathcal{N}}\,\Phi]\!]$. Also note that $[\![\vec{\rho}\,\mathtt{false}[\Phi]]\!] = \mathtt{lfp}_Z([\![\vec{\mathcal{N}}\,\mathtt{false}]\!] \cup ([\![\vec{\mathcal{N}}\,(\Phi \wedge Z)]\!])) = \emptyset$, and again analogously $[\![\bar{\rho}\,\mathtt{false}[\Phi]]\!] = \emptyset$.

**Proposition 4.** *Let $\mathcal{T}$ be a model, $s \in S$ a point, and $\Phi$ a SL formula. Then $s \models \Phi$ iff $s \in [\![\Phi]\!]$.*

*Proof.* The proof is immediate for all operators except reachability. Consider e.g. the next operator: we have that $s \models \vec{\mathcal{N}}\,\Phi$ iff $s_1 \models \Phi$ for some $s_1 \in R(s)$ iff $R(s) \cap [\![\Phi]\!] \neq \emptyset$, the latter by inductive hypothesis. And we noted in Remark 3 that the semantics of the derived operators is respected, i.e. $[\![\vec{\rho}\,\Phi[\mathtt{false}]]\!] = [\![\vec{\mathcal{N}}\,\Phi]\!]$.

Now, recall that by Lemma 2 $s \models \vec{\rho}\Phi_1[\Phi_2]$ iff $s \models \vec{\mathcal{N}}\,\Phi_1 \vee \vec{\mathcal{N}}\,(\Phi_2 \wedge \vec{\rho}\Phi_1[\Phi_2])$.

($\Longrightarrow$) By induction on the length of the path $ss_1 \ldots s_n$ verifying $s \models \vec{\rho}\Phi_1[\Phi_2]$. If $n = 1$, then $s_1 \models \Phi_1$, hence $s_1 \in [\![\Phi_1]\!]$ and $s \in [\![\vec{\mathcal{N}}\,\Phi_1]\!]$, Otherwise, $s_1 \models \Phi_2 \wedge \vec{\rho}\Phi_1[\Phi_2]$ with a path of length $n - 1$, hence $s_1 \in [\![\Phi_2 \wedge \vec{\rho}\Phi_1[\Phi_2]]\!]$ and $s \in [\![\vec{\mathcal{N}}\,(\Phi_2 \wedge \vec{\rho}\Phi_1[\Phi_2])]\!]$. In both cases, we have that $s \in [\![\vec{\rho}\,\Phi_1[\Phi_2]]\!]$.

($\Longleftarrow$) By induction on the number $r$ of recursive steps $Z_1, Z_2 \ldots Z_r$. If $r = 1$, then $s \in [\![\vec{\mathcal{N}}\,\Phi_1]\!]$, hence there exists $s_1 \in R(S) \cap [\![\Phi_1]\!]$, thus $s_1 \in R(S)$ and $s_1 \models [\![\Phi_1]\!]$. For $r = n + 1$, either $s \in [\![\vec{\mathcal{N}}\,\Phi_1]\!]$, and we fall back to the previous case, or $s \in [\![\vec{\mathcal{N}}\,(\Phi_2 \wedge Z_n)]\!]$. Hence there exists $s_1 \in R(S) \cap [\![\Phi_2]\!] \cap [\![Z_n]\!]$, so the by inductive hypothesis $s_1 \models \Phi_2 \wedge \vec{\rho}\Phi_1[\Phi_2]$. In both cases, we have that $s \in \vec{\rho}\,\Phi_1[\Phi_2]$.

## 3.2   SL vs. CTL

We make here precise the connection between SL and CTL. The state formulas for the existential fragment of CTL (ECTL) can be expressed by the grammar

$$\Psi ::= \mathtt{true} \mid a \mid \neg\Psi \mid \Psi \wedge \Psi \mid \exists O\Psi \mid \exists U(\Psi, \Psi)$$

Note that this fragment is not as expressive as CTL, since it is missing the operators $\forall O\Psi$ and $\forall U(\Psi, \Psi)$. And while the former is CTL-equivalent to $\neg\exists O\neg\Psi$, the latter cannot be expressed in the fragment: it requires the operator $\exists\Box$.

Let us now prove the equivalence of ECTL with the forward fragment of SL (FSL), i.e. SL without the backward operator $\bar{\rho}$. We do not recall here the semantics for CTL, and we refer the reader to a standard reference such as [3].

*The Encodings.* For any FSL formula $\Phi$ we must obtain an ECTL formula $[\![\Phi]\!]$ such that for any model $\mathcal{T}$ and state $s$ in $\mathcal{T}$ we have that $s \models_{SL} \Phi$ iff $s \models_{CTL} [\![\Phi]\!]$. Clearly, the Boolean operators are mapped one-to-one, while $\vec{\rho}\,\Phi_1[\Phi_2]$ is mapped into $\exists O(\exists U([\![\Phi_2]\!], [\![\Phi_1]\!]))$. Note that, as a derived operator, $\vec{\mathcal{N}}\Phi$ is mapped into $\exists O(\exists U(\texttt{false}, [\![\Phi]\!]))$, which is CTL-equivalent to $\exists O[\![\Phi]\!]$.

Viceversa, for any ECTL formula $\Psi$ we must obtain a FSL formula $\|\Psi\|$. As before, the Boolean operators are mapped one-to-one, while instead $\exists O\Psi$ is mapped to $\vec{\mathcal{N}}\|\Psi\|$ and $\exists U(\Psi_1, \Psi_2)$ is mapped to $\|\Psi_2\| \vee (\|\Psi_1\| \wedge \vec{\rho}\,\|\Psi_2\|[\![\|\Psi_1\|]\!])$. Again, for any model $\mathcal{T}$ and state $s$ in $\mathcal{T}$ we have that $s \models_{CTL} \Psi$ iff $s \models_{SL} \|\Psi\|$.

*Encodings are Mutually Inverse.* We proceed by structural induction, assuming that for the sub-formulae it holds that $[\![\|\Psi\|]\!]$ and $\Psi$ are CTL-equivalent and $\|[\![\Phi]\!]\|$ and $\Phi$ are SL-equivalent.

Starting from ECTL, we have that

- $[\![\|\exists O\Psi\|]\!] = [\![\vec{\mathcal{N}}\|\Psi\|]\!] = \exists O(\exists U(\texttt{false}, [\![\|\Psi\|]\!]))$
- $[\![\|\exists U(\Psi_1, \Psi_2)\|]\!] = [\![\|\Psi_2\| \vee (\|\Psi_1\| \wedge \vec{\rho}\,\|\Psi_2\|[\![\|\Psi_1\|]\!])]\!] = [\![\|\Psi_2\|]\!] \vee ([\![\|\Psi_1\|]\!] \wedge [\![\vec{\rho}\,\|\Psi_2\|[\![\|\Psi_1\|]\!]]\!]) = [\![\|\Psi_2\|]\!] \vee ([\![\|\Psi_1\|]\!] \wedge \exists O(\exists U([\![\|\Psi_1\|]\!], [\![\|\Psi_2\|]\!])))$

and the result follows since for the former case $\exists U(\texttt{false}, [\![\|\Psi\|]\!])$ is CTL-equivalent to $[\![\|\Psi\|]\!]$ and for the latter case it is the well-known expansion law for $\exists U$.

Moving from FSL, we have that

- $\|[\![\vec{\rho}\,\Phi_1[\Phi_2]\!]]\!]\| = \|\exists O(\exists U([\![\Phi_2]\!], [\![\Phi_1]\!]))\| = \vec{\mathcal{N}}\|\exists U([\![\Phi_2]\!], [\![\Phi_1]\!])\| = \vec{\mathcal{N}}([\![\|\Phi_1\|]\!] \vee ([\![\|\Phi_2\|]\!] \wedge \vec{\rho}\,[\![\|\Phi_1\|]\!][\![[\![\|\Phi_2\|]\!]]\!])$

The two formulae are SL-equivalent, as shown in Lemma 2.

## 4 Quantified Spatial Logics

We now move to a Quantified Spatial Logic (QSL). In the following, we fix a set of typed variables $V = V_P \uplus V_S$ ranged over by $x, y, x_P, y_P, x_S, y_S \dots$

**Definition 7.** *The formulae $\Phi$ of QSL are given by the grammar*

$$\Phi ::= \texttt{true} \mid a \mid x \mid x = y \mid \neg\Phi \mid \Phi \wedge \Phi \mid \vec{\rho}\,\Phi[\Phi] \mid \overleftarrow{\rho}\,\Phi[\Phi] \mid \exists_x.\Phi$$

**Definition 8.** *Let $\mathcal{T}$ be a model. The semantics of a QSL formula $\Phi$ with respect to a point $s \in S$ and a substitution $\eta : V \rightharpoonup P \uplus S$ is given by the rules*

- $s, \eta \models x_P$ *if* $s \in L(\eta(x_P))$
- $s, \eta \models x_S$ *if* $s = \eta(x_S)$
- $s, \eta \models x = y$ *if* $\eta(x) = \eta(y)$
- $s, \eta \models \exists_{x_P}.\Phi$ *if there exists a proposition $a_1$ such that $s, \eta[{}^{a_1}/_{x_P}] \models \Phi$*
- $s, \eta \models \exists_{x_S}.\Phi$ *if there exists a point $s_1$ such that $s, \eta[{}^{s_1}/_{x_S}] \models \Phi$*

*for $\eta[{}^{a_1}/_{x_P}]$ and $\eta[{}^{s_1}/_{x_S}]$ the standard extensions of a substitution $\eta$.*

For the sake of readability, we showed only the rules for the variables and the existential operators, and implicitly assumed that equality $x = y$ is well-typed.

*Remark 4.* Variables may take values either in points or in atomic propositions. Hence, we have statements such as $s, \eta \models x \wedge y$ with $\eta(x)$ a point and $\eta(y)$ an atomic proposition, which still has a clear semantics: it holds if $s = \eta(x)$ and $s \in L(\eta(y))$. As recalled, we implicitly have typed equality $x =_\tau y$ for variables $x, y$ of the same type $\tau$, which is either $S$ for points or $P$ for atomic propositions. With respect to [11], we lack an explicit constant `this` for characterising the current state, which can be obtained by using a point variable $x$ occurring in a formula $\Phi$ and simply checking $s \models \exists_x.(x \wedge \Phi)$. In general, the equality $x_S = a$, meaning that the point associated to $x_S$ by a substitution $\eta$ satisfies proposition $a$, is recovered as $x_S \wedge a$. Also lacking are equalities $x_P = a$ for proposition $a$: they seem less relevant, and could be added with little effort.

*Remark 5.* A further step along the lines above is to assume that variables take values in sets of points, i.e. $\eta : V \to 2^S$, obtaining a second-order quantification. It would simply mean to add an additional type for second-order variables and possibly a monadic operator $\in$, as in $x \in X$. Note that in this case the equality $x = y$ for point variables could be derived as $\forall_X. x \in X \iff y \in X$.

### 4.1  Denotational Semantics for QSL

The denotational meaning of a QSL formula $\Phi$ is going to be a set of points in our model $\mathcal{T}$. We define our denotational mapping $[\![\cdot]\!]_\eta$ as follows.

**Definition 9.** *Let $\mathcal{T}$ be a model. The denotational semantics of a QSL formula $\Phi$ with respect to a substitution $\eta$ is given by the rules*

- $[\![x_P]\!]_\eta = L(\eta(x_P))$
- $[\![x_S]\!]_\eta = \{\eta(x_S)\}$
- $[\![x = y]\!]_\eta = \begin{cases} S \text{ if } \eta(x) = \eta(y) \\ \emptyset \text{ otherwise} \end{cases}$
- $[\![\exists_{x_P}.\Phi]\!]_\eta = \bigcup_{a \in P} [\![\Phi]\!]_{\eta[a/x_P]}$
- $[\![\exists_{x_S}.\Phi]\!]_\eta = \bigcup_{s \in S} [\![\Phi]\!]_{\eta[s/x_S]}$

As before, we just showed the rules for variables and existential operators.

*Remark 6.* It should be no surprise now that the equality $[\![\exists_x.\vec{\mathcal{N}}\Phi]\!]_\eta = [\![\vec{\mathcal{N}}\exists_x.\Phi]\!]_\eta$ holds for any $\eta$, and similarly for $\vec{\mathcal{N}}$. Indeed, the shape of a single frame never changes, hence QSL satisfies what is called the domain-preserving property.

Now, let $\bot : V \rightharpoonup P \uplus S$ denote the always undefined substitution.

**Proposition 5.** *Let $\mathcal{T}$ be a model, $s \in S$ a point, and $\Phi$ a closed QSL formula. Then $s, \bot \models \Phi$ iff $s \in [\![\Phi]\!]_\bot$.*

*Remark 7.* Quantification over atomic proposition is intended to model the idea of quantifying over "labels" that identify sets of points sharing similar features, in such a way that the number of available labels is infinite and model-dependent. This does not imply that the set of labels that are *present* in each state is infinite:

it could as well be that, in a system with infinite states, the number of labels of *each state* is finite, but no state has the same set of labels. In this situation, typical e.g. of nominal computations [23], it might not be possible to know in advance which labels will be present in a state of the model. But this does not rule out the possibility of asking meaningful questions, such as "is there a point labelled with $x_P$ in the current state, which in the next state will *not* be labelled with $x_P$ and *near* to a point labelled with $x_P$?", which could be interpreted as the entity denoted by $x_P$ has moved by one step in one instant of time.

Although it is perhaps easier to grasp the intuition when models have a temporal aspect, the idea is also useful in purely spatial situations. One case often occurring in computational imaging is that of reasoning about *connected components*. Consider a spatial formula $\phi$ interpreted over a digital image. No matter what $\phi$ is, the semantics will identify the set of points $S$ on which $\phi$ holds. In many situations one could be interested in questions such as "identify the set of points $S'$ that belong to a *connected region* $R$ of $S$, which also satisfies $\psi$". In our view, connectedness is not a primitive of the logical language (as connectedness is just one example of application of quantification over atomic propositions!). Rather, the *model* must contain enough information to reason – in this case – about connected components, by having a different atomic proposition for each component[1]. In this situation, one does not know in advance neither how many components (hence, atomic propositions) will be available, nor the exact set of labels, but still, existential quantification over atomic propositions can be used.

*Example 2.* Using the aforementioned encoding of connected component labels as atomic propositions, we are able to identify entities in a given space. Continuing from Example 1, we now assume that for each frame the set of atomic properties includes colours as well as the labels of the connected components of the yellow pixels. We can now characterise in each frame the pixels on the border of the active Pac-Man as $\Phi = yellow \wedge \forall x_P.\, (\vec{\rho}\,(x_P \wedge yellow)[black] \implies x_P)$, since the active Pac-Man cannot reach those outside while these latter are mutually reachable, and the whole active Pac-Man via the formula $yellow \wedge \vec{\rho}\, \Phi[yellow]$.

## 5    Spatio-temporal Logics

The definitions below have the following rationale. In analysing video frames we basically deal with sequences of graphs, each one of them a snapshot of an image. The structure of the graph remains the same: only the labelling changes, i.e. the atomic propositions each point satisfies. Also, note that when we state properties of sequences of graphs, we often do not even have a way to generate such sequences. Think e.g. about the scans of the brain: they are given by physicians,

---

[1] In model checking, this is accomplished at model definition time, by including a *non-logical* operator which performs a *labelling of connected components*, taking as input a Boolean-labelled frame and returning a integer-labelled frame, where each connected component is identified by a unique integer. See [10] where the on-GPU variant of the spatial model checker VoxLogicA has been endowed with such a primitive.

and they are not obtained by a set of rules, since they are just snapshots taken at certain intervals of time. We might thus have a single trace as model. This is the reason for the choice of linear time, hence of our Spatio-Temporal Logic (STL): the following proposals could be easily rephrased in terms of computational trees.

**Definition 10.** *The formulae $\Phi$ of STL are given by the grammar*

$$\Phi ::= \mathtt{true} \mid a \mid \neg\Phi \mid \Phi \wedge \Phi \mid \vec{\rho}\,\Phi[\Phi] \mid \overleftarrow{\rho}\,\Phi[\Phi] \mid \mathsf{O}\Phi \mid \mathsf{U}(\Phi,\Phi)$$

A spatio-temporal model $\mathcal{S}$ is a four-tuple $\langle S, P, R, \Lambda_0 \rangle$, where $S$ is a set of points, $P$ a set of atomic propositions, $R : S \to 2^S$ a (spatial) relation, $\Lambda_0 \subset \Lambda^+$ a set of temporal traces of length at least 1, for $\Lambda = \{L \mid L : P \to 2^S\}$ the set of labelings. We give the semantics of the formulae with respect to a point $s$ and a finite trace $\lambda$. Given a temporal trace $\lambda = L_0 L_1 \ldots, L_n$, we denote by $\lambda(i)$ the sequence $L_i L_{i+1} \ldots$, by $\lambda_i$ its $i$-th component $L_i$, and with $l(\lambda)$ its length $n+1$.

**Definition 11.** *Let $\mathcal{T}$ be a spatio-temporal model. The semantics of a STL formula $\Phi$ with respect to a point $s \in S$ and a temporal trace $\lambda \in \Lambda_0$ is given by the rules*

- $s, \lambda \models \mathsf{O}\Phi$ *if* $1 < l(\lambda)$ *and* $s, \lambda(1) \models \Phi$
- $s, \lambda \models \mathsf{U}(\Phi_1, \Phi_2)$ *if there exists* $k < l(\lambda)$ *such that* $s, \lambda(k) \models \Phi_2$ *and* $s, \lambda(j) \models \Phi_1$ *for all* $j = 0 \ldots k - 1$

*Remark 8.* Since we are using *finite* temporal traces, a few considerations are in order. As a start, a formula $\mathsf{O}\Phi$ is satisfiable by a temporal trace if it is of length at least two, so that $\mathtt{last} = \neg\mathsf{O}\mathtt{true}$ actually characterises its last component. Such an operator allows an easy characterisation for the nesting of temporal operators, since $\Box\Diamond\Phi$ and $\Diamond\Box\Phi$ are equivalent to $\Diamond(\mathtt{last} \wedge \Phi)$ [19].

A related question is which axioms hold. As an example, $\neg\mathsf{O}\Phi$ and $\mathsf{O}\neg\Phi$ are equivalent only for temporal traces of length at least two, since $\mathsf{O}\Phi$ is always false for temporal traces of length 1. Instead, the usual unfolding axiom for the until operator holds, that is, $s, \lambda \models \mathsf{U}(\Phi_1, \Phi_2)$ iff $s, \lambda \models \Phi_2 \vee (\Phi_1 \wedge \mathsf{OU}(\Phi_1, \Phi_2))$.

The interaction between spatial and temporal operators needs to be explored. For example, $\vec{\rho}\,\mathsf{O}a[\mathsf{O}b]$ is equivalent to $\mathsf{O}(\vec{\rho}\,a[b])$, since the structure of the model (points and their relations) never changes during the steps of a temporal trace.

## 5.1   Denotational Semantics of STL

The denotational meaning of a formula $\Phi$ is going to be a set of points in our model $\mathcal{T}$. We define our denotational mapping $[\![\cdot]\!]_\lambda$ as follows.

**Definition 12.** *Let $\mathcal{T}$ be a spatio-temporal model. The denotational semantics of a STL formula $\Phi$ with respect to a temporal trace $\lambda \in \Lambda_0$ is given by the rules*

- $[\![\mathsf{O}\Phi]\!]_\lambda = \begin{cases} [\![\Phi]\!]_{\lambda(1)} & \textit{if } 1 < l(\lambda) \\ \emptyset & \textit{otherwise} \end{cases}$
- $[\![\mathsf{U}(\Phi_1, \Phi_2)]\!]_\lambda = \mathtt{lfp}_W\left([\![\Phi_2]\!]_\lambda \cup ([\![\Phi_1]\!]_\lambda \cap [\![\mathsf{O}\,W]\!]_\lambda)\right)$

As before, we presented the mapping only for the newly introduced temporal operators. As for the reachability operators, the fix-point for U is well-defined.

**Proposition 6.** *Let $\mathcal{T}$ be a spatio-temporal model, $s \in S$ a point, $\lambda \in \Lambda_0$ a temporal trace, and $\Phi$ a STL formula. Then $s, \lambda \models \Phi$ iff $s \in \llbracket \Phi \rrbracket_\lambda$.*

*Proof.* Similarly to the operators of spatial logics in the proof of Proposition 4, we will basically proceed by induction on the structure of the formulae, considering here also the length of the temporal trace. We just look at the additional temporal operators, noting that it is obvious for the next operator $O\Phi$. Recall, see Remark 8, that formulae $U(\Phi_1, \Phi_2)$ and $\Phi_2 \vee (\Phi_1 \wedge OU(\Phi_1, \Phi_2))$ are equivalent.

($\Longleftarrow$) By induction on the structure of the formulae and the length of the temporal trace. If $s, \lambda \models \Phi_2 \vee (\Phi_1 \wedge OU(\Phi_1, \Phi_2))$, then either $s, \lambda \models \Phi_2$, hence $s \in \llbracket \Phi_2 \rrbracket_\lambda$ by inductive hypothesis, or $s, \lambda \models OU(\Phi_1, \Phi_2)$, thus $s \models \Phi_1$ and $s, \lambda(1) \models U(\Phi_1, \Phi_2)$, hence $s \in \llbracket \Phi_1 \rrbracket_\lambda \cap \llbracket O\,W \rrbracket_\lambda$ by inductive hypothesis.

($\Longrightarrow$) By induction on the number $r$ of recursive steps $W_1, W_2 \ldots W_r$ and the length of the temporal trace. If $r = 1$, then $s \in \llbracket \Phi_2 \rrbracket_\lambda$, and we are done by inductive hypothesis. For $r = n + 1$, we have that either $s \in \llbracket \Phi_2 \rrbracket_\lambda$, and we fall back to the previous case, or $s \in \llbracket \Phi_2 \rrbracket \cap \llbracket OW_n \rrbracket_\lambda$, and in particular $s \in \llbracket W_n \rrbracket_{\lambda(1)}$, Thus by inductive hypothesis $s, \lambda \models \Phi_2 \wedge OU(\Phi_1, \Phi_2))$.

# 6 All Together Now

Recall that with our logics we aim to state properties about the single snapshots of a sequence, detailing their changes along time. The Quantified Spatio-Temporal Logic (QSTL) is obtained just by the combination of all the operators introduced so far, thus quantifying "globally" along the whole length of a trace.

**Definition 13.** *The formulae $\Phi$ of QSTL are given by the grammar*

$$\Phi ::= \texttt{true} \mid a \mid x \mid x = y \mid \neg\Phi \mid \Phi \wedge \Phi \mid \vec{\rho}\,\Phi[\Phi] \mid \overleftarrow{\rho}\,\Phi[\Phi] \mid O\Phi \mid U(\Phi, \Phi) \mid \exists_x.\Phi$$

**Definition 14.** *Let $\mathcal{T}$ be a spatio-temporal model. The semantics of a QSTL formula $\Phi$ with respect to a point $s \in S$, a substitution $\eta : V \rightharpoonup P \uplus S$, and a temporal trace $\lambda \in \Lambda_0$ is given by the rules*

- $s, \eta, \lambda \models \texttt{true}$
- $s, \eta, \lambda \models a$ *if* $a \in \lambda_0(s)$
- $s, \eta, \lambda \models x_P$ *if* $s \in \lambda_0(\eta(x_P))$
- $s, \eta, \lambda \models x_S$ *if* $s = \eta(x_S)$
- $s, \eta, \lambda \models x = y$ *if* $\eta(x) = \eta(y)$
- $s, \eta, \lambda \models \neg\Phi$ *if* $s, \eta, \lambda \not\models \Phi$
- $s, \eta, \lambda \models \Phi_1 \wedge \Phi_2$ *if* $s, \eta, \lambda \models \Phi_1$ *and* $s, \eta, \lambda \models \Phi_2$
- $s, \eta, \lambda \models \vec{\rho}\,\Phi_1[\Phi_2]$ *if there exists a spatial path $ss_1 \ldots s_n$ in $\mathcal{T}$ such that $s_n, \eta, \lambda \models \Phi_1$ and $s_j, \eta, \lambda \models \Phi_2$ for all $j = 1 \ldots n - 1$*
- $s, \eta, \lambda \models \overleftarrow{\rho}\,\Phi_1[\Phi_2]$ *if there exists a spatial path $s_0 \ldots s_{n-1}s$ in $\mathcal{T}$ such that $s_0, \eta, \lambda \models \Phi_1$ and $s_j, \eta, \lambda \models \Phi_2$ for all $j = 1 \ldots n - 1$*

- $s, \eta, \lambda \models \mathsf{O}\varPhi$ *if* $1 < l(\lambda)$ *and* $s, \eta, \lambda(1) \models \varPhi$
- $s, \eta, \lambda \models \mathsf{U}(\varPhi_1, \varPhi_2)$ *if there exists* $k < l(\lambda)$ *such that* $s, \eta, \lambda(k) \models \varPhi_2$ *and* $s, \eta, \lambda(j) \models \varPhi_1$ *for all* $j = 0 \ldots k-1$
- $s, \eta, \lambda \models \exists_{x_P}.\varPhi$ *if there exists a proposition* $a_1$ *such that* $s, \eta[^{a_1}/_x], \lambda \models \varPhi$
- $s, \eta, \lambda \models \exists_{x_S}.\varPhi$ *if there exists a point* $s_1$ *such that* $s, \eta[^{s_1}/_x], \lambda \models \varPhi$

We can now combine the denotational mappings seen before to get $[\![\cdot]\!]_{\eta,\lambda}$, and to finally obtain our concluding result.

**Proposition 7.** *Let* $\mathcal{T}$ *be a spatio-temporal model,* $s \in S$ *a point,* $\lambda \in \Lambda_0$ *a temporal trace, and* $\varPhi$ *a QSTL formula. Then* $s, \perp, \lambda \models \varPhi$ *iff* $s \in [\![\varPhi]\!]_{\perp,\lambda}$.

*Example 3.* We shall now discuss a scenario where all the features of the language are needed. This example is aimed at tracking the identity of objects along the temporal axis. As said in Remark 7, quantifiers on atomic propositions are used to assign labels in order to identify entities, being these points or regions. In Example 2 these labels represent connected components. In this case, instead, we assume that, for each ghost, the spatio-temporal model encodes the identity of each "lifespan" (the time between a character first appears on the screen, and the moment it is caught, or the game finishes) via a unique atomic proposition. In other terms, for each ghost and each lifespan, a separate atomic proposition always identifies all the pixels that the ghost occupies on screen.

We use this idea to define a logic formula $\phi$ that is true at the pixels of the orange ghost, in the current state, if and only if such ghost will be caught by Pac-Man in a subsequent state. We shall use the derived operator "somewhere" defined as $\mathcal{F}\phi = \bar{\rho}\phi[\mathbf{true}]$. We define the formula *orange* $\wedge \exists x_P.x_P \wedge \mathsf{U}(true, \mathcal{F}(x_P \wedge blue \wedge \bar{\mathcal{N}}pacman))$. Note that, if the formula is true at a point $s$, then that point is orange, and there is an atomic proposition $x_P$ which holds in $s$, thus, by construction, it represents the identity of the current ghost. Furthermore, by definition of $\mathsf{U}$, such atomic proposition is still true at *some* point $s'$ of the space, in some future state, with $s'$ in contact with a point of Pac-Man, which entails that the ghost is caught in the same sense of Example 1.

## 7   Conclusions and Future Works

We developed a quantified spatio-temporal logic, and showed how this can be used to state spatial properties, possibly involving the identity of individuals, in models that evolve along time. The logic thus represents a significant improvement in expressivity with respect to SLCS [16]. Differently from [11], we adopted linear time operators and an operational semantics based on finite traces. We also introduced a denotational semantics and proved its equivalence with the operational one. Despite its simplicity, the Pac-Man example clarifies the usefulness of the logic in applicative domains such as video stream analysis and lesion tracking in medical imaging.

Concerning future works, we plan to investigate decidability and axiomatisations of the logic. Bisimilarity and minimisation of models can be also of interest, akin to the work for SLCS in [15]. As far as applications are concerned, we will aim at developing a prototype spatial model checker combining temporal and existential operators, and to use it in medical imaging case studies.

# A   Some Hints from Quantified Modal Algebras

This appendix recalls basic notions of (quantified) modal and conjugate algebras, which inspired the way we provided our logics with a denotational semantics.

## A.1   Boolean and Modal Algebra

We recall the basics of boolean and modal algebras and discuss some axioms.

**Definition 15.** *A Boolean algebra $\mathcal{A}$ is a 6-tuple $\langle A, \vee, 0, \wedge, 1, \neg \rangle$ such that the triples $\langle A, \vee, 0 \rangle$ and $\langle A, \wedge, 1 \rangle$ are ACI (associative, commutative and with identity) monoids satisfying the usual distributivity and negation rules.*

The usual negation rule means that $a \vee \neg a = 1$ and $a \wedge \neg a = 0$. A Boolean algebra is equivalently described as a *complemented distributive lattice*. In particular $a \vee b = a$ iff $a \wedge b = b$ and $a \leq b$ iff $\neg b \leq \neg a$. The partial order on $A$ is induced by $a \leq b$ if $a \vee b = b$, so that 0 is bottom and 1 is top. A well-known example of such a structure is the boolean algebra of powersets of a set, that gives rise to the algebra $\mathcal{A} = \langle \mathcal{P}(A), \cup, \emptyset, \cap, A, {}^c \rangle$. We say that a Boolean algebra $\mathcal{A}$ is *complete* if every subset of $A$ has a least upper bound (LUB).

**Definition 16.** *A modal algebra $\mathcal{M}$ is a 7-tuple $\langle A, \vee, 0, \wedge, 1, \neg, \Diamond \rangle$ such that the 6-tuple $\langle A, \vee, 0, \wedge, 1, \neg \rangle$ is a Boolean algebra and $\Diamond : A \to A$ is a function satisfying $\Diamond 0 = 0$ and $\Diamond(a \vee b) = \Diamond a \vee \Diamond b$.*
*A modal algebra is complete if the underlying Boolean algebra is complete and $\Diamond(\bigvee_i a_i) = \bigvee_i \Diamond a_i$ for any $i$.*

Monotonicity of $\Diamond$ is implied by the second axiom, which also yields that $\Diamond 1 = 1$. If $\mathcal{M}$ is finite (i.e. the set $A$ is finite), then $\mathcal{M}$ is obviously complete.
We define the usual derived operator $\Box a = \neg \Diamond \neg a$. Note that $\Box 1 = 1$, $\Box a \wedge b = \Box a \wedge \Box b$, and $\Box$ is monotone with respect to the induced partial order

*Remark 9.* Modal algebras provide denotational models for propositional modal logics. Assuming a semantical function $[\cdot]$ mapping a formula into an element of the modal algebra chosen as model, the formula $\phi$ is valid in the logics if $[\phi] = 1$. Also, note that $[\phi \implies \rho] = 1$ is equivalent to prove that $[\phi] \leq [\rho]$, assuming that $[\cdot]$ preserves the operators $\neg$ and $\vee$ (hence, all the operators).
It is immediate that the axiom $K$, i.e. $\Box(\phi \implies \rho) \implies (\Box\phi \implies \Box\rho)$, holds in any modal algebra. By Boolean manipulation the formula is equivalent to $(\Box\phi \wedge (\Box(\phi \implies \rho)) \implies \Box\rho$. Hence, it suffices to prove that in a modal algebra it holds $(\Box a \wedge \Box(a \implies b)) \leq \Box b$. Due to the distributivity of $\Box$, this is equivalent to prove that $\Box(a \wedge b) \leq \Box b$, which holds by monotonicity.
Also, note that what is called the necessitation rule for modal logics based on $K$ holds, since $a = 1$ implies $\Box a = \Box 1 = 1$.

**Definition 17.** *Let $M$ be a modal algebra whose partial order is $\leq$. Its necessity and iteration axioms are $M = a \leq \Diamond a$, $4 = \Diamond\Diamond a \leq \Diamond a$, and $B = a \leq \Box\Diamond a$.*

Axioms are given in terms of the $\Diamond$ operator, but they can be rewritten using the $\Box$ operator, with the reversed inequality. Hence, $M$ and $4$ can be equivalently expressed in terms of $\Box$ as $\Box a \leq a$ and $\Box a \leq \Box\Box a$, respectively, as well as $B$ is equivalent to $\Diamond\Box a \leq a$. Note that assuming $M$ and $4$ implies that $\Diamond\Diamond a = \Diamond a$.

*Remark 10.* Axioms $M$, $4$, and $B$ are known as reflexivity, transitivity, and symmetry axioms, respectively, since for modal algebras arising from Kripke frames those are the properties imposed on the underlying relation [18]. Modal algebras satisfying $M$ and $4$ are called closure algebras and are models of $S4$, while those satisfying all three axioms are called monadic algebras and are models of $S5$.

## A.2　Quantified Modal Algebras

While modal algebras represent models for propositional modal logics, moving to first order quantification require the introduction of *cylindric operators*, a well-known abstraction for existential quantifiers [21].

**Cylindric Operators.** We fix a Boolean algebra $\mathcal{A}$ and a set of variables $V$.

**Definition 18 (cylindric Boolean algebras).** *A cylindric operator $\exists$ over $\mathcal{A}$ and $V$ is a family of monotone operators $\exists_x : A \to A$ indexed by elements in $V$ such that for all $a, b \in A$ and $x, y \in V$ it holds $a \leq \exists_x a$, $\exists_x \exists_y a = \exists_y \exists_x a$, and $\exists_x(a \wedge \exists_x b) = \exists_x a \wedge \exists_x b$.*
　　*Let $a \in A$. The* support *of $a$ is the set of variables $sv(a) = \{x \mid \exists_x a \neq a\}$.*

An element of the algebra stands for a formula possibly containing free variables. We restrict our attention to elements $a$ with finite support, i.e., such that $sv(a)$ is finite: this means that $a$ is a formula containing a finite set of variables.
　　Now we fix a modal algebra $\mathcal{M}$ with underlying Boolean algebra $\mathcal{A}$.

**Definition 19 (cylindric modal algebras).** *A cylindric operator $\exists$ over $\mathcal{M}$ and $V$ is a cylindric operator over $\mathcal{A}$ and $V$ such that for all $a \in A$ and $x \in V$ it holds $\exists_x \Diamond a = \Diamond \exists_x a$.*

*Remark 11.* The inequalities $\exists_x \Diamond a \geq \Diamond \exists_x a$ and $\exists_x \Diamond a \leq \Diamond \exists_x a$ are known as *Barcan formula* and *converse Barcan formula* in the literature [6]. The axiom in Definition 19 is thus only one of the possible choices, and it boils down to require what is called "domain preservation", namely, the domain is preserved along the evolution. Instead, $\exists_x \Diamond a \leq \Diamond \exists_x a$ witnesses a possible domain restriction, while analogously we may have a domain increase with the reverse $\exists_x \Diamond a \geq \Diamond \exists_x a$.
　　The axiom implies $sv(\Diamond a) \subseteq sv(a)$, since $\exists_x a = a$ implies $\exists_x \Diamond a = \Diamond \exists_x a = \Diamond a$.

**Soft Modal Algebras.** We now show how to build a modal algebra that admits cylindric operators. Let us fix a modal algebra $\mathcal{M}$ with underlying Boolean algebra $\mathcal{A}$ and a set of variables $V$.

**Proposition 8.** *Let $D$ be a set of elements, $F$ the set of functions $\eta : V \to D$, and $\Gamma$ the set of functions $\gamma : F \to A$. The 7-tuple $\mathcal{F} = \langle \Gamma, \vee, 0, \wedge, 1, \neg, \Diamond \rangle$ is a modal algebra, whose operators and constants are lifted from $\mathcal{M}$. If $\mathcal{M}$ is complete, so is $\mathcal{F}$.*

For example, $0$ in $\mathcal{F}$ is the function such that $0(\eta) = 0$ for all $\eta$, and so on. In particular, note that $\gamma_1 \leq \gamma_2$ means that $\gamma_1(\eta) \leq \gamma_2(\eta)$ for all $\eta$.

Let us now additionally fix a set $D$, and given $\eta : V \to D$, we denote as $\eta[^d/_x]$ the function coinciding with $\eta$ except for $x$, where $\eta[^d/_x](x) = d$.

**Proposition 9.** *Let $D$ be finite. The cylindric operator $\exists$ over $\mathcal{F}$ and $V$ is defined as $(\exists_x \gamma)(\eta) = \bigvee_{d \in D} \gamma(\eta[^d/_x])$.*

If $\mathcal{M}$ is complete, the finiteness of $D$ can be dropped.

*Remark 12.* By definition, $\exists_x \gamma = \gamma$ means that for all $\eta$ we have $\bigvee_{d \in D} \gamma(\eta[^d/_x]) = \gamma(\eta)$, which is equivalent to say that for all $d$ we have $\gamma(\eta[^d/_x]) = \gamma(\eta)$. Intuitively, if $\gamma$ represents a formula possibly containing free variables, $x$ cannot be among them. Conversely, $x \in sv(\gamma)$ if there is a function $\eta$ and elements $b, c \in D$ such that $\gamma(\eta[^b/_x]) \neq \gamma(\eta[^c/_x])$, intuitively meaning that $x$ does occur free in $\gamma$.

## A.3    Conjugate Modal Algebras

Algebras that employ more than one modal operator are said to be *multimodal.* We focus here on a particular kind of such algebras, called *conjugate algebras.*

**Definition 20.** *A conjugate algebra $\mathcal{D}$ is a 8-tuple $\langle A, \vee, 0, \wedge, 1, \neg, \Diamond_1, \Diamond_2 \rangle$ such that both 7-tuples $\langle A, \vee, 0, \wedge, 1, \neg, \Diamond_1 \rangle$ and $\langle A, \vee, 0, \wedge, 1, \neg, \Diamond_2 \rangle$ are modal algebras and moreover it holds $a \leq \Box_1 \Diamond_2 a \wedge \Box_2 \Diamond_1 a$.*
*A conjugate algebra is* complete *if both the underlying modal algebras are so.*

What is noteworthy is a well-known characterisation via just the $\Diamond$ operators.

**Lemma 3.** *$\mathcal{D}$ is a conjugate algebra iff it holds $\Diamond_1 a \wedge b = 0 \Leftrightarrow a \wedge \Diamond_2 b = 0$.*

*Remark 13.* The lemma is stated by using the more standard notion of the axiom on $\Diamond$. An alternative, friendlier version is $\Diamond_1 a \leq b \Leftrightarrow a \leq \Box_2 b$. The proof of the equivalence between the two axioms is straightforward, and it exploits the following law holding in Boolean algebras, namely $a \wedge b = 0$ iff $a \leq \neg b$.

# References

1. Aiello, M., Pratt-Hartmann, I., van Benthem, J.: Handbook of Spatial Logics. Springer, Dordrecht (2007). https://doi.org/10.1007/978-1-4020-5587-4
2. Awodey, S., Kishida, K.: Topology and modality: the topological interpretation of first-order modal logic. Rev. Symbolic Logic **1**(2), 146–166 (2008)
3. Baier, C., Katoen, J.P.: Principles of Model Checking. The MIT Press (2008)
4. Banci Buonamici, F., Belmonte, G., Ciancia, V., Latella, D., Massink, M.: Spatial logics and model checking for medical imaging. Softw. Tools Technol. Transf. **22**(2), 195–217 (2020)
5. Bartocci, E., Gol, E., Haghighi, I., Belta, C.: A formal methods approach to pattern recognition and synthesis in reaction diffusion networks. IEEE Trans. Control Netw. Syst. **5**(1), 308–320 (2016)
6. Basin, D., Matthews, S., Viganò, L.: Labelled modal logics: quantifiers. J. Logic Lang. Inform. **7**(3), 237–263 (1998)
7. Belmonte, G., Broccia, G., Ciancia, V., Latella, D., Massink, M.: Feasibility of spatial model checking for nevus segmentation. In: Bliudze, S., Gnesi, S., Plat, N., Semini, L. (eds.) FormaliSE@ICSE 2021, pp. 1–12. IEEE (2021)
8. Belmonte, G., Ciancia, V., Latella, D., Massink, M.: VoxLogicA: a spatial model checker for declarative image analysis. In: Vojnar, T., Zhang, L. (eds.) TACAS 2019. LNCS, vol. 11427, pp. 281–298. Springer, Cham (2019). https://doi.org/10.1007/978-3-030-17462-0_16
9. Bennett, B., Cohn, A., Wolter, F., Zakharyaschev, M.: Multi-dimensional modal logic as a framework for spatio-temporal reasoning. Appl. Intell. **17**(3), 239–251 (2002)
10. Bussi, L., Ciancia, V., Gadducci, F.: Towards a spatial model checker on GPU. In: Peters, K., Willemse, T.A.C. (eds.) FORTE 2021. LNCS, vol. 12719, pp. 188–196. Springer, Cham (2021). https://doi.org/10.1007/978-3-030-78089-0_12
11. Bussi, L., Ciancia, V., Gadducci, F., Latella, D., Massink, M.: On binding in the spatial logics for closure spaces. In: Margaria, T., Steffen, B. (eds.) Leveraging Applications of Formal Methods, Verification and Validation. Verification Principles, ISoLA 2022. LNCS, vol. 13701, pp. 479–497. Springer, Cham (2022). https://doi.org/10.1007/978-3-031-19849-6_27
12. Bussi, L., Ciancia, V., Gadducci, F., Latella, D., Massink, M.: Towards model checking video streams using VoxLogicA on GPUs. In: Bowles, J., Broccia, G., Pellungrini, R. (eds.) From Data to Models and Back, DataMod 2021. LNCS, vol. 13268, pp. 78–90. Springer, Cham (2022). https://doi.org/10.1007/978-3-031-16011-0_6
13. Ciancia, V., Latella, D., Massink, M., Paškauskas, R., Vandin, A.: A tool-chain for statistical spatio-temporal model checking of bike sharing systems. In: Margaria, T., Steffen, B. (eds.) ISoLA 2016. LNCS, vol. 9952, pp. 657–673. Springer, Cham (2016). https://doi.org/10.1007/978-3-319-47166-2_46
14. Ciancia, V., Gilmore, S., Grilletti, G., Latella, D., Loreti, M., Massink, M.: Spatio-temporal model checking of vehicular movement in public transport systems. Softw. Tools Technol. Transf. **20**(3), 289–311 (2018)
15. Ciancia, V., Groote, J.F., Latella, D., Massink, M., de Vink, E.P.: Minimisation of spatial models using branching bisimilarity. In: Chechik, M., Katoen, J.P., Leucker, M. (eds.) Formal Methods, FM 2023. LNCS, vol. 14000, pp. 263–281. Springer, Cham (2023). https://doi.org/10.1007/978-3-031-27481-7_16

16. Ciancia, V., Latella, D., Loreti, M., Massink, M.: Specifying and verifying properties of space. In: Diaz, J., Lanese, I., Sangiorgi, D. (eds.) TCS 2014. LNCS, vol. 8705, pp. 222–235. Springer, Heidelberg (2014). https://doi.org/10.1007/978-3-662-44602-7_18

17. Gabelaia, D., Kontchakov, R., Kurucz, A., Wolter, F., Zakharyaschev, M.: Combining spatial and temporal logics: expressiveness vs. complexity. Artif. Intell. Res. **23**, 167–243 (2005)

18. Garson, J.: Modal logic. In: Zalta, E.N., Nodelman, U. (eds.) The Stanford Encyclopedia of Philosophy (Spring 2023 Edition). Metaphysics Research Lab, Stanford University (2023)

19. Giacomo, G.D., Vardi, M.Y.: Linear temporal logic and linear dynamic logic on finite traces. In: Rossi, F. (ed.) IJCAI 2013, pp. 854–860. IJCAI/AAAI (2013)

20. Kishida, K.: Neighborhood-sheaf semantics for first-order modal logic. In: van Ditmarsch, H., Fernández-Duque, D., Goranko, V., Jamroga, W., Ojeda-Aciego, M. (eds.) M4M/LAMAS 2011. ENTCS, vol. 278, pp. 129–143. Elsevier (2011)

21. Monk, J.D.: An introduction to cylindric set algebras. Log. J. IGPL **8**(4), 451–496 (2000)

22. Nenzi, L., Bartocci, E., Bortolussi, L., Loreti, M.: A logic for monitoring dynamic networks of spatially-distributed cyber-physical systems. Logical Meth. Comput. Sci. **18**(1), 4:1–4:30 (2022)

23. Pitts, A.M.: Nominal Sets: Names and Symmetry in Computer Science. Cambridge University Press (2013)

24. Tsigkanos, C., Kehrer, T., Ghezzi, C.: Modeling and verification of evolving cyber-physical spaces. In: Bodden, E., Schäfer, W., van Deursen, A., Zisman, A. (eds.) ESEC/FSE 2017, pp. 38–48. ACM (2017)

# Logic of the Hide and Seek Game: Characterization, Axiomatization, Decidability

Qian Chen[1,2] and Dazhu Li[3,4(✉)]

[1] The Tsinghua -UvA Joint Research Centre for Logic, Department of Philosophy, Tsinghua University, Beijing, China
`chenq21@mails.tsinghua.edu.cn`
[2] Institute for Logic, Language and Computation, University of Amsterdam, Amsterdam, Netherlands
[3] Institute of Philosophy, Chinese Academy of Sciences, Beijing, China
[4] Department of Philosophy, University of Chinese Academy of Sciences, Beijing, China
`lidazhu@ucas.ac.cn`

**Abstract.** The logic of the hide and seek game LHS was proposed to reason about search missions and interactions between agents in pursuit-evasion environments. As proved in [15,16], having an equality constant in the language of LHS drastically increases its computational complexity: the satisfiability problem for LHS with multiple relations is undecidable. In this work, we improve the existing proof for the undecidability by showing that LHS with a single relation is undecidable. With the findings of [15,16], we provide a van Benthem style characterization theorem for the expressive power of the logic. Finally, by 'splitting' the language of LHS⁻, a crucial fragment of LHS without the equality constant, into two 'isolated parts', we provide a complete Hilbert style proof system for LHS⁻ and prove that its satisfiability problem is decidable, whose proofs would indicate significant differences between the proposals of LHS⁻ and of ordinary product logics. Although LHS and LHS⁻ are frameworks for interactions of 2 agents, all results in the article can be easily transferred to their generalizations for settings with any $n > 2$ agents.

**Keywords:** Logic of the hide and seek game · Axiomatization · Modal logic · Expressive power · Decidability

## 1 Introduction

The logic LHS of the hide and seek game was introduced in [2] that promotes a study of graph game design in tandem with matching modal logics, and then was probed in [15] and its further extension [16]. The logic provides us with a platform to reason about search problems and interactions between agents with entangled goals, as in the case of the hide and seek game [2] (or the game of

N. Gierasimczuk and F. R. Velázquez-Quesada (Eds.): DaLí 2023, LNCS 14401, pp. 20–34, 2024.
https://doi.org/10.1007/978-3-031-51777-8_2

cops and robber [17]): in a fixed graph, two players Hider and Seeker take turns to move to a successor of their own positions, and Seeker tries to move to the same position with Hider while Hider aims to avoid Seeker.

To describe the game, the language of LHS contains two modalities for the movements of the two players and a constant $I$ expressing that the positions of Hider and Seeker are the same. Semantically, models for LHS are the same as relational models for basic modal logic [4], while formulas are evaluated at *two* states, which intuitively represent the positions of the two players.

In addition to the applications to the graph games, LHS is also of interest from other perspectives. One of them is that the framework links up the study of graph game logics with many other important fields: as illustrated in [15,16], LHS and *its fragment* LHS⁻ *without the constant $I$* have close connection with *product logics*, including K × K [8] and its extension K ×$^\delta$ K with *a diagonal constant $\delta$* [9,11,12]; the framework LHS is highly relevant to *cylindric modal logics* that also contain constants for equality [21]; and both K ×$^\delta$ K and cylindric modal logics in turn provide a link between LHS and *cylindric algebra* proposed in [10]. Moreover, the framework LHS provides an instance showing how an innocent looking proposal $I$ for equality can drastically increase the computational complexity of the logic: as proved in [15,16], the satisfiability problem for LHS with multiple binary relations is undecidable.

In this work, we will explore the further properties of LHS and LHS⁻. First, we improve the existing undecidability proof for LHS with multiple relations and show that LHS with a single relation is undecidable (Sect. 3). Then, based on the notions of first-order translation and bisimulations for LHS given in [16], we develop a van Benthem style characterization theorem for the expressiveness of LHS (Sect. 4). Next, for LHS⁻, we develop a complete Hilbert style calculus and show that its satisfiability problem is decidable (Sect. 5), and our proofs would indicate important differences between the proposals of LHS⁻ and K × K. Also, we discuss related work and point out a few lines of further study (Sect. 6). It is instructive to notice that although LHS and LHS⁻ are frameworks for the hide and seek game with 2 players, all these results can be transferred to the logics generalizing LHS and LHS⁻ for the settings with $n > 2$ players, but we stick to discussing the systems LHS and LHS⁻ for simplicity.

## 2   Basics of the Logic of the Hide and Seek Game

We start by concisely introducing the basics of LHS, including its language and semantics, and providing preliminary observations on its properties.

**Definition 1.** *Let* $\mathsf{L} = \{p_i^l : i \in \mathbb{N}\}$ *and* $\mathsf{R} = \{p_i^r : i \in \mathbb{N}\}$ *be two disjoint countable sets of propositional variables. The* language $\mathcal{L}$ *of* LHS *is given by:*

$$\mathcal{L} \ni \varphi ::= p^l \mid p^r \mid I \mid \neg\varphi \mid \varphi \wedge \varphi \mid \Box\varphi \mid \blacksquare\varphi,$$

*where* $p^l \in \mathsf{L}$ *and* $p^r \in \mathsf{R}$.

Abbreviations $\top, \bot, \lor, \rightarrow$ are as usual, and we use $\Diamond, \blacklozenge$ for the dual operators of $\Box$ and $\blacksquare$ respectively. For convenience, we call $\Box$ and $\Diamond$ 'white modalities' and call $\blacksquare$ and $\blacklozenge$ 'black modalities'. Also, the notion of *subformulas* is as usual, and for any $\varphi \in \mathcal{L}$, we employ $\mathsf{Sub}(\varphi)$ for *the set of subformulas of* $\varphi$. In what follows, we use $\mathcal{L}^-$ for the part of $\mathcal{L}$ without $I$, which is the language for $\mathsf{LHS}^-$.

A *frame* is a tuple $\mathfrak{F} = (W, R)$ such that $W$ is a non-empty set of states and $R \subseteq W \times W$ is a binary relation on $W$. A *model* $\mathfrak{M} = (W, R, V)$ equips a frame with a valuation function $V : \mathsf{L} \cup \mathsf{R} \rightarrow \mathcal{P}(W).$[1] For any $s, t \in W$, we call $\langle \mathfrak{M}, s, t \rangle$ a *pointed* $\mathsf{LHS}$-*model*. For simplicity, we usually write $\mathfrak{M}, s, t$ for it. For each $w \in W$ and $U \subseteq W$, we define $R(w) = \{v \in W : Rwv\}$ and $R(U) = \bigcup_{u \in U} R(u)$.

**Definition 2.** *Let* $\mathfrak{M} = (W, R, V)$ *be a model and* $s, t \in W$. *Truth of formulas* $\varphi \in \mathcal{L}$ *at* $\langle \mathfrak{M}, s, t \rangle$, *written as* $\mathfrak{M}, s, t \models \varphi$, *is defined recursively as follows:*

$$
\begin{aligned}
\mathfrak{M}, s, t \models p^l &\Leftrightarrow s \in V(p^l) \\
\mathfrak{M}, s, t \models p^r &\Leftrightarrow t \in V(p^r) \\
\mathfrak{M}, s, t \models I &\Leftrightarrow s = t \\
\mathfrak{M}, s, t \models \neg\varphi &\Leftrightarrow \mathfrak{M}, s, t \not\models \varphi \\
\mathfrak{M}, s, t \models \varphi \land \psi &\Leftrightarrow \mathfrak{M}, s, t \models \varphi \text{ and } \mathfrak{M}, s, t \models \psi \\
\mathfrak{M}, s, t \models \Box\varphi &\Leftrightarrow \mathfrak{M}, s', t \models \varphi \text{ for all } s' \in R(s) \\
\mathfrak{M}, s, t \models \blacksquare\varphi &\Leftrightarrow \mathfrak{M}, s, t' \models \varphi \text{ for all } t' \in R(t)
\end{aligned}
$$

Notions of *satisfiability*, *validity* and *logical consequence* are defined in the usual manner. Let $\mathsf{LHS}$ denote the set of all valid formulas.

*Remark 1.* With the semantic clause for $I$, we can see that it is essentially a proposal to capture equality. Similarly, @-*operators* in ordinary *hybrid logics* are also proposals for equality (see e.g., [4, Chapter 7.3]). For discussion on differences between these two approaches, see [19]. Also, in $\mathsf{LHS}$, $\Box\varphi$ and $\blacksquare\varphi$ move along a common relation $R$, but it is also interesting to consider the case that models contain two different relations, one for each player, which means that Hider and Seeker can make different moves. We believe that the results developed in the article can be transferred to this variant by adapting our proofs.

Let $\mathfrak{M} = (W, R, V)$ be a model and $U \subseteq W$. We say *a model* $\mathfrak{M}' = (W', R', V')$ *is generated from* $\mathfrak{M} = (W, R, V)$ *by* $U$, if $\mathfrak{M}'$ is the smallest model satisfying the following: $U \subseteq W'$, $R(W') \subseteq W'$, $R' = R \cap (W' \times W')$, and for each $p \in \mathsf{L} \cup \mathsf{R}$, $V'(p) = V(p) \cap W'$.

**Proposition 1.** *Let* $\mathfrak{M}' = (W', R', V')$ *be a submodel of* $\mathfrak{M} = (W, R, V)$ *generated by* $\{s, t\} \subseteq W$. *For any formula* $\varphi \in \mathcal{L}$, $\mathfrak{M}, s, t \models \varphi$ *iff* $\mathfrak{M}', s, t \models \varphi$.

*Proof.* It goes by induction on formulas. We omit the details to save space. $\square$

---

[1] For any set $A$, we use $\mathcal{P}(A)$ for its *power set*.

# 3   Undecidability of LHS

As stated, [15,16] proved that the satisfiability problem for LHS with multiple binary relations is undecidable. In this part, we show that LHS with a single relation is undecidable as well, which is an improvement of the existing proof.

**Theorem 1.** *The satisfiability problem for* LHS *is undecidable.*

We show this by reduction of the $\mathbb{N} \times \mathbb{N}$ tiling problem [18] to the satisfiability problem for LHS. Let $\mathbb{T} = \{T_1, \ldots, T_n\}$ be some fixed set of tile types. For each $T_i \in \mathbb{T}$, we use up$(T_i)$, down$(T_i)$, left$(T_i)$ and right$(T_i)$ to represent the colors of its up, down, left and right edges, respectively. We say that $\mathbb{T}$ *tiles* $\mathbb{N} \times \mathbb{N}$ if there is a function $g : \mathbb{N} \times \mathbb{N} \to \mathbb{T}$ such that for all $n, m \in \mathbb{N}$,

$$\text{right}(g(n, m)) = \text{left}(g(n + 1, m)) \text{ and } \text{up}(g(n, m)) = \text{down}(g(n, m + 1)).$$

Functions satisfying the conditions above are called *tiling functions*. In what follows, to show that LHS is undecidable, we present a formula $\varphi_{\mathbb{T}}$ such that

$$\varphi_{\mathbb{T}} \text{ is satisfiable if and only if } \mathbb{T} \text{ tiles } \mathbb{N} \times \mathbb{N}.$$

Let Label $= \{u, r\} \cup \{t_i : 1 \le i \le n\}$ be a set of labels. Let $\mathsf{NV}^L = \{p^l : p \in \mathsf{Label}\}$ and $\mathsf{NV}^R = \{p^r : p \in \mathsf{Label}\}$ be sets of new variables. For convenience, we denote $\bigvee_{1 \le i \le n} t_i^l$ by $t^l$ and $\bigvee_{1 \le i \le n} t_i^r$ by $t^r$. We write $\Diamond_u \varphi$ for $t^l \wedge \Diamond(u^l \wedge \Diamond(t^l \wedge \varphi))$ and $\Diamond_r \varphi$ for $t^l \wedge \Diamond(r^l \wedge \Diamond(t^l \wedge \varphi))$. Operators $\blacklozenge_u$ and $\blacklozenge_r$ are defined similarly. The dual of these operators are defined as usual, for example, $\Box_u \varphi := \neg \Diamond_u \neg \varphi$.

The formula $\varphi_{\mathbb{T}}$ is the conjunction of those in the groups below. To facilitate discussion, let $\mathfrak{M} = (W, R, V)$ be a model and $w, v \in W$ s.t. $\mathfrak{M}, w, v \models \varphi_{\mathbb{T}}$.

**Group 1 (Basic requirements):**

    (SP) $I \wedge \Box\Box\blacklozenge I \wedge \Diamond t^l$
    (VL1) $\Box\blacksquare(I \to \bigwedge_{p \in \mathsf{Label}}(p^l \leftrightarrow p^r))$
    (VL2) $\Box \bigwedge_{p \in \mathsf{Label}}(p^l \leftrightarrow \bigwedge_{p \ne q \in \mathsf{Label}} \neg q^l)$

Let us explain the meanings of the formulas in Group 1. Intuitively, we can treat $t, u, r$ as labels. The formula (SP) says that $w = v$, $R(R(w)) \subseteq R(w)$ and there is some $v \in R(w)$ which is labelled by $t$. (VL1) indicates that for any $s \in R(w)$, its 'left-label' and 'right-label' are always the same. Moreover, (VL2) shows that every point $s \in R(w)$ has exactly one label.

**Group 2 (Grid requirements):**

    (TU1) $\Box\blacksquare(t^l \wedge I \to \Diamond(u^l \wedge \blacksquare(u^r \to I)))$
    (TU2) $\Box\blacksquare(u^l \wedge I \to \Diamond(t^l \wedge \blacksquare(t^r \to I)))$
    (TR1) $\Box\blacksquare(t^l \wedge I \to \Diamond(r^l \wedge \blacksquare(r^r \to I)))$
    (TR2) $\Box\blacksquare(r^l \wedge I \to \Diamond(t^l \wedge \blacksquare(t^r \to I)))$
    (URT) $\Box\blacksquare(t^l \wedge I \to \Box_u\blacksquare_r\Diamond_r\blacklozenge_u I)$

We can assume that $\mathfrak{M}$ is a model generated by $w \in W$ (Proposition 1). Let

$$R_u = \{\langle s, t \rangle \in R : \mathfrak{M}, s, t \models t^l \wedge t^r \text{ and for some } x \in V(u^l), \ sRx \text{ and } xRt\}.$$

It follows from (TU1) and (TU2) that for all $s \in R(w)$, $|R_u(s)| = 1$.[2] Similarly, we can define $R_r$, and by (TR1) and (TR2), for all $s \in R(w)$, $|R_r(s)| = 1$. From (URT), we can infer that for all $v \in R(w)$, $R_r(R_u(v)) = R_u(R_r(v))$.

**Group 3 (Tiling the model):**

(T1) $\Box(t^l \rightarrow \bigwedge_{i=1}^{n}(t_i^l \rightarrow \Diamond_u \bigvee_{1 \leq j \leq n \ \& \ \text{up}(T_i)=\text{down}(T_j)} t_j^l))$;
(T2) $\Box(t^l \rightarrow \bigwedge_{i=1}^{n}(t_i^l \rightarrow \Diamond_r \bigvee_{1 \leq j \leq n \ \& \ \text{right}(T_i)=\text{left}(T_j)} t_j^l))$.

The formulas in Group 3 are standard, which tell us that $\mathbb{T}$ 'tiles' $R(w) \cap V(t^l)$.

**Lemma 1.** *If $\mathbb{T}$ tiles $\mathbb{N} \times \mathbb{N}$, then $\varphi_{\mathbb{T}}$ is satisfiable.*

*Proof.* Let $h : \mathbb{N} \times \mathbb{N} \rightarrow \mathbb{T}$ be a tiling function. Define $\mathfrak{M}_h = (W, R, V)$ as follows:

- $W = W_0 \cup \{s\}$, where $W_0 = \{\langle n, m \rangle \in \mathbb{N} \times \mathbb{N} : n \times m \text{ is even}\}$
- $R = R_r \cup R_u \cup (\{s\} \times W_0)$, where
    - $R_r = \{\langle \langle k, 2l \rangle, \langle k+1, 2l \rangle \rangle : k, l \in \mathbb{N}\}$
    - $R_u = \{\langle \langle 2k, l \rangle, \langle 2k, l+1 \rangle \rangle : k, l \in \mathbb{N}\}$
- $V$ is a valuation such that
    - $V(r^l) = V(r^r) = \{\langle 2k+1, 2l \rangle \in W : k, l \in \mathbb{N}\}$
    - $V(u^l) = V(u^r) = \{\langle 2k, 2l+1 \rangle \in W : k, l \in \mathbb{N}\}$
    - $V(t_i^l) = V(t_i^r) = \{\langle 2k, 2l \rangle \in W : k, l \in \mathbb{N}, h(k, l) = T_i\}$ for all $1 \leq i \leq n$.
    - $V(p^l) = V(q^r) = \emptyset$ for all other $p^l, q^r \in \mathsf{L} \cup \mathsf{R}$.

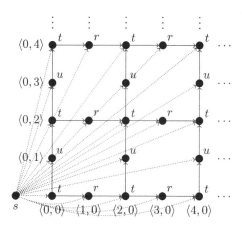

**Fig. 1.** The model $\mathfrak{M}_h$: Both dotted arrows and solid arrows represent the relation $R$.

The model $\mathfrak{M}_h$ is shown in Fig. 1. It is easy to verify that $\mathfrak{M}_h, s, s \models \varphi_{\mathbb{T}}$.     □

**Lemma 2.** *If $\varphi_{\mathbb{T}}$ is satisfiable, then $\mathbb{T}$ tiles $\mathbb{N} \times \mathbb{N}$.*

*Proof.* Suppose $\mathfrak{M} = (W, R, V)$ is a model generated by $s \in W$ and $\mathfrak{M}, s, s \models \varphi_{\mathbb{T}}$ (Proposition 1). It suffices to define a tiling function $g : \mathbb{N} \times \mathbb{N} \to T$. By (SP1), $V(t^l) \neq \emptyset$. Also, it follows from (TU1) and (TU2) that for each $w \in V(t^l)$, there is exactly one state $v \in V(t^l)$ such that $wRxRv$ for some $x \in V(u^l)$, and we denote the state $v$ by $\mathsf{up}(w)$. Then $\mathsf{up} : V(t^l) \to V(t^l)$ is a function. Similarly, due to (TR1) and (TR2), we can define a function $\mathsf{right} : V(t^l) \to V(t^l)$. Let $w_0 \in V(t^l)$. We inductively define a function $g : \mathbb{N} \times \mathbb{N} \to V(t^l)$ as follows:

$$g(\langle 0, 0 \rangle) = w_0, \quad g(\langle n, m + 1 \rangle) = \mathsf{up}(g(\langle n, m \rangle)), \quad g(\langle n + 1, m \rangle) = \mathsf{right}(g(\langle n, m \rangle)).$$

Now from (URT) we can infer that for each $w \in V(t^l)$, $\mathsf{up}(\mathsf{right}(w)) = \mathsf{right}(\mathsf{up}(w))$. Then, for all $\langle n, m \rangle \in \mathbb{N} \times \mathbb{N}$,

$$\mathsf{up}(g(\langle n + 1, m \rangle)) = \mathsf{up}(\mathsf{right}(g(\langle n, m \rangle))) = \mathsf{right}(\mathsf{up}(g(\langle n, m \rangle))) = \mathsf{right}(g(\langle n, m + 1 \rangle)).$$

Hence function $g$ is well-defined. Let $h : V(t^l) \to T$ be the function such that for each $1 \leq i \leq n$, $h(w) = T_i$ if and only if $w \in V(t^l_i)$. Finally, by the formulas in Group 3, it is clear that $h \circ g : \mathbb{N} \times \mathbb{N} \to T$ is a tiling function.     □

## 4     van Benthem Characterization Theorem

This section is devoted to the expressive power of LHS. Precisely, based on the notions of its first-order translation and bisimulations developed in [15,16], we will provide a van Benthem style characterization theorem for the logic.

Let $\mathcal{L}^1$ be the first-order language consisting of the following: a countable set $\mathsf{P} = \{P^l_i, P^r_i : i \in \mathbb{N}\}$ of unary predicates, a binary relation $R$ and equality $\equiv$. For any two variables $x$ and $y$, the first-order translation $\mathrm{T}_{\langle x, y \rangle} : \mathcal{L} \to \mathcal{L}^1$ for LHS is given recursively as follows:

$$\mathrm{T}_{\langle x, y \rangle}(p^l_i) := P^l_i x \qquad \mathrm{T}_{\langle x, y \rangle}(p^r_i) := P^r_i y \qquad \mathrm{T}_{\langle x, y \rangle}(I) := (x \equiv y)$$

$$\mathrm{T}_{\langle x, y \rangle}(\neg\varphi) := \neg\mathrm{T}_{\langle x, y \rangle}(\varphi) \qquad \mathrm{T}_{\langle x, y \rangle}(\varphi \wedge \psi) := \mathrm{T}_{\langle x, y \rangle}(\varphi) \wedge \mathrm{T}_{\langle x, y \rangle}(\psi)$$

$$\mathrm{T}_{\langle x, y \rangle}(\Box\varphi) := \forall z(Rxz \to \mathrm{T}_{\langle z, y \rangle}(\varphi)) \qquad \mathrm{T}_{\langle x, y \rangle}(\blacksquare\varphi) := \forall z(Ryz \to \mathrm{T}_{\langle x, z \rangle}(\varphi))$$

For any set $\Phi \subseteq \mathcal{L}$ of formulas, we define $\mathrm{T}_{\langle x, y \rangle}(\Phi) := \{\mathrm{T}_{\langle x, y \rangle}(\varphi) : \varphi \in \Phi\}$. Now, the following result indicates the correctness of the translation:

**Proposition 2 ([16]).** *For any pointed LHS-model $\langle \mathfrak{M}, s, t \rangle$ and $\varphi \in \mathcal{L}$,*

$$\mathfrak{M}, s, t \models \varphi \text{ if and only if } \mathfrak{M} \models \mathrm{T}_{\langle x, y \rangle}(\varphi)[s, t].^3$$

Let us recall the notions of LHS-*bisimulation* and LHS-*saturation* in [16]:

---

[3] By $\mathfrak{M} \models \mathrm{T}_{\langle x, y \rangle}(\varphi)[s, t]$, we mean that when values of $x, y$ in $\mathrm{T}_{\langle x, y \rangle}(\varphi)$ are $s, t$ respectively, $\mathrm{T}_{\langle x, y \rangle}(\varphi)$ is satisfied by $\mathfrak{M}$.

**Definition 3** ([15,16]). *Let $\mathfrak{M} = (W, R, V)$ and $\mathfrak{M}' = (W', R', V')$ be models. A binary relation $Z \subseteq (W \times W) \times (W' \times W')$ is called an* LHS-*bisimulation between $\mathfrak{M}$ and $\mathfrak{M}'$, notation $Z : \mathfrak{M} \leftrightarrow \mathfrak{M}'$, if the following conditions hold for all $s, t, v \in W$ and $s', t', v' \in W'$:*

- *If $\langle s, t \rangle Z \langle s', t' \rangle$, then for all $p \in \mathsf{L} \cup \mathsf{R}$, $\mathfrak{M}, s, t \models p$ if and only if $\mathfrak{M}', s', t' \models p$.*
- *If $\langle s, t \rangle Z \langle s', t' \rangle$ and $v \in R(s)$, then there is $v' \in R'(s')$ s.t. $\langle v, t \rangle Z \langle v', t' \rangle$.*
- *If $\langle s, t \rangle Z \langle s', t' \rangle$ and $v \in R(t)$, then there is $v' \in R'(t')$ s.t. $\langle s, v \rangle Z \langle s', v' \rangle$.*
- *If $\langle s, t \rangle Z \langle s', t' \rangle$ and $v' \in R'(s')$, then there is $v \in R(s)$ s.t. $\langle v, t \rangle Z \langle v', t' \rangle$.*
- *If $\langle s, t \rangle Z \langle s', t' \rangle$ and $v' \in R'(t')$, then there is $v \in R(t)$ s.t. $\langle s, v \rangle Z \langle s', v' \rangle$.*
- *If $\langle s, t \rangle Z \langle s', t' \rangle$, then $s = t$ if and only if $s' = t'$.*

If there is an LHS-bisimulation $Z : \mathfrak{M} \leftrightarrow \mathfrak{M}'$ s.t. $\langle s, t \rangle Z \langle s', t' \rangle$, then we say that $\langle \mathfrak{M}, s, t \rangle$ *is* LHS-*bisimular to* $\langle \mathfrak{M}', s', t' \rangle$ and write $\langle \mathfrak{M}, s, t \rangle \leftrightarrow \langle \mathfrak{M}', s', t' \rangle$.

**Definition 4** ([15,16]). *Let $\mathfrak{M} = (W, R, V)$ be a model. A set $\Delta$ of formulas is said to be satisfiable in $X \subseteq W \times W$ if $\mathfrak{M}, s, t \models \Delta$ for some $\langle s, t \rangle \in X$. Then $\mathfrak{M}$ is said to be* LHS-*saturated if for all $\Phi \subseteq \mathcal{L}$ and $w, v \in W$:*

1. *If every finite subset $\Sigma$ of $\Phi$ is satisfiable in $R(w) \times \{v\}$, then $\Phi$ is satisfiable in $R(w) \times \{v\}$.*
2. *If every finite subset $\Sigma$ of $\Phi$ is satisfiable in $\{w\} \times R(v)$, then $\Phi$ is satisfiable in $\{w\} \times R(v)$.*

**Proposition 3** ([15,16]). *If $\langle \mathfrak{M}, s, t \rangle \leftrightarrow \langle \mathfrak{M}', s', t' \rangle$, then $\langle \mathfrak{M}, s, t \rangle$ and $\langle \mathfrak{M}', s', t' \rangle$ satisfy the same* LHS-*formulas.*

Proposition 3 indicates that LHS-bisimulation given above is what we desired. The converse of Proposition 3 holds for LHS-*saturated* models:

**Proposition 4** ([15,16]). *For all $\mathfrak{M}$ and $\mathfrak{M}'$ that are* LHS-*saturated, if $\langle \mathfrak{M}, s, t \rangle$ and $\langle \mathfrak{M}', s', t' \rangle$ satisfy the same formulas of* LHS, *then $\langle \mathfrak{M}, s, t \rangle \leftrightarrow \langle \mathfrak{M}', s', t' \rangle$.*

Let $\Gamma(x_1, \ldots, x_n) \subseteq \mathcal{L}^1$. We say that $\mathfrak{M} = (W, R, V)$ *realizes* $\Gamma$ if there are $a_1, \ldots, a_n \in W$ s.t. $\mathfrak{M} \models \gamma[a_1, \ldots, a_n]$ for all $\gamma \in \Gamma$. Also, let $A \subseteq W$. For each $a \in A$, let $c_a$ be a constant symbol. Let $\mathcal{L}_A^1 = \mathcal{L}^1 \cup \{c_a : a \in A\}$ and let $\mathfrak{M}_A$ denote the $\mathcal{L}_A^1$-expansion $\mathfrak{M}$ such that for all $a \in A$, $a$ has the value $c_a$.

**Definition 5.** *A model $\mathfrak{M} = (W, R, V)$ is $\omega$-saturated, if for all $A \subseteq W$, $\mathfrak{M}_A$ realizes every $\Gamma(x) \subseteq \mathcal{L}_A^1$ whose finite subsets are all realized in $\mathfrak{M}_A$.*

**Proposition 5.** *All $\omega$-saturated models $\mathfrak{M} = (W, R, V)$ are* LHS-*saturated.*

*Proof.* Let $\Sigma \subseteq \mathcal{L}$ be finitely satisfiable in $R(w) \times \{v\}$ and $w, v \in W$. Then let $\Delta(x) = \{Rc_w x\} \cup \{\mathrm{T}_{\langle x, y \rangle}(\varphi)[y/c_v] : \varphi \in \Sigma\}$. Every finite subset of $\Delta(x)$ is realized by some $u \in R(w)$ in $\mathfrak{M}_{\{w, v\}}$ (Proposition 2). Since $\mathfrak{M}$ is $\omega$-saturated, $\Delta(x)$ is realized in $\mathfrak{M}_{\{w, v\}}$. So, there is some $u \in W$ such that $\mathfrak{M}_{\{w, v\}} \models \Delta(x)[u]$. Thus $u \in R(w)$ and $\langle \mathfrak{M}, u, v \rangle \models \Sigma$. Similarly, if $\Sigma$ is finitely satisfiable in $\{w\} \times R(v)$, then $\Sigma$ is satisfiable in $\{w\} \times R(v)$. □

**Corollary 1.** *Let* $\mathfrak{M} = (W, R, V)$ *and* $\mathfrak{M}' = (W', R', V')$ *be* $\omega$*-saturated models,* $s, t \in W$ *and* $s', t' \in W'$. *If* $\langle \mathfrak{M}, s, t \rangle$ *and* $\langle \mathfrak{M}', s', t' \rangle$ *satisfy the same* $\mathcal{L}$*-formulas, then* $\langle \mathfrak{M}, s, t \rangle \leftrightarroweq \langle \mathfrak{M}', s', t' \rangle$.

Let $\mathfrak{M}$ be a model, $\mathbb{I}$ a countable set and $U$ an incomplete ultrafilter over $\mathbb{I}$. Then we write $\prod_U \mathfrak{M}$ for the ultrapower of $\mathfrak{M}$ modulo $U$.[4]

**Proposition 6.** *Let* $\mathfrak{M} = (W, R, V)$ *be a model,* $\mathbb{I}$ *a countable set and* $U$ *an incomplete ultrafilter over* $\mathbb{I}$. *For each* $w \in W$, *let* $f_w = \mathbb{I} \times \{w\}$. *Then,*

1. $\prod_U \mathfrak{M}$ *is* $\omega$*-saturated.*
2. *For any* $\alpha(x, y) \in \mathcal{L}^1$ *and* $s, t \in W$, $\mathfrak{M} \models \alpha[s, t]$ *iff* $\prod_U \mathfrak{M} \models \alpha[(f_s)_U, (f_t)_U]$.
3. *For any* $\mathcal{L}$*-formula* $\varphi$ *and* $s, t \in W$, $\mathfrak{M}, s, t \models \varphi$ *iff* $\prod_U \mathfrak{M}, (f_s)_U, (f_t)_U \models \varphi$.

*Proof.* The first item follows from [5, p.384, Theorem 6.1.1]. The second follows from [5, p.217, Theorem 4.1.9]. The last one follows from the second item and Proposition 2 immediately.  □

We say that *an* $\mathcal{L}^1$*-formula* $\alpha(x, y)$ *is invariant for* LHS*-bisimulation*, if for all $\langle \mathfrak{M}, s, t \rangle$ and $\langle \mathfrak{M}', s', t' \rangle$ that are LHS-bisimilar, $\mathfrak{M} \models \alpha[s, t]$ iff $\mathfrak{M}' \models \alpha[s', t']$. Now we can provide a van Benthem style characterization theorem for LHS:

**Theorem 2.** *For any* $\alpha(x, y) \in \mathcal{L}^1$, $\alpha(x, y)$ *is invariant for* LHS*-bisimulation if and only if* $\models \alpha \leftrightarrow \beta$ *for some* $\beta(x, y) \in \mathrm{T}_{\langle x, y \rangle}(\mathcal{L})$.

*Proof.* The right-to-left direction is easy, which follows straightforward from Proposition 3. For the other direction, let $\alpha(x, y) \in \mathcal{L}^1$ be invariant for LHS-bisimulation. Define $\mathrm{modal}(\alpha) := \{\beta \in \mathrm{T}_{\langle x, y \rangle}(\mathcal{L}) : \alpha \models \beta\}$. We show $\mathrm{modal}(\alpha) \models \alpha$. Let $\mathfrak{M}$ be a model such that $\mathfrak{M} \models \mathrm{modal}(\alpha)[a, b]$. It suffices to show that $\mathfrak{M} \models \alpha[a, b]$. Let $\Phi$ be a set of formulas defined by

$$\Phi := \{\beta(x, y) \in \mathrm{T}_{\langle x, y \rangle}(\mathcal{L}) : \mathfrak{M} \models \beta[a, b]\} \cup \{\alpha(x, y)\}.$$

We claim that $\Phi$ is satisfiable. Suppose $\Phi$ is not satisfiable. Then there is a finite $\Phi_0 \subseteq \Phi$ with $\Phi_0 \models \neg\alpha$, which entails $\alpha \models \neg\bigwedge \Phi_0$ and so $\mathfrak{M} \models \neg\bigwedge \Phi_0[a, b]$. Note that $\bigwedge \Phi_0 \in \Phi$, we see $\mathfrak{M} \models \bigwedge \Phi_0[a, b]$, which is a contradiction. Thus, $\Phi$ is satisfiable and there is a model $\mathfrak{N}$ and states $w, u$ with $\mathfrak{N} \models \Phi[w, u]$. Then by Proposition 2, $\langle \mathfrak{M}, a, b \rangle$ and $\langle \mathfrak{N}, w, u \rangle$ satisfy the same $\mathcal{L}$-formulas. Let $U$ be an incomplete ultrafilter over $\mathbb{N}$. Then by Proposition 6(3), $\langle \prod_U \mathfrak{M}, (f_a)_U, (f_b)_U \rangle$ and $\langle \prod_U \mathfrak{N}, (f_c)_U, (f_d)_U \rangle$ satisfy the same $\mathcal{L}$-formulas. By Proposition 6(1) and Corollary 1, $\langle \prod_U \mathfrak{M}, (f_a)_U, (f_b)_U \rangle \leftrightarroweq \langle \prod_U \mathfrak{N}, (f_w)_U, (f_u)_U \rangle$. Since $\mathfrak{N} \models \alpha[w, u]$, by Proposition 6(2), $\prod_U \mathfrak{N} \models \alpha[(f_w)_U, (f_u)_U]$. Since $\alpha(x, y)$ is invariant for LHS-bisimulation, we have $\prod_U \mathfrak{M} \models \alpha[(f_a)_U, (f_b)_U]$. By Proposition 6(2), $\mathfrak{M} \models \alpha[a, b]$. Hence, $\mathrm{modal}(\alpha) \models \alpha$. By the Compactness Theorem, there is a finite $\Sigma \subseteq \mathrm{modal}(\varphi)$ such that $\Sigma \models \alpha$. Then we see $\models \alpha \leftrightarrow \bigwedge \Sigma$.  □

Finally, it is worthwhile to notice that when we restrict our attention to LHS⁻, by adapting the arguments for LHS, we can also obtain a characterization theorem for the expressiveness of LHS⁻, but we omit the details to save space.

---

[4] For the definitions of *ultrafilter* and *ultrapower of models*, see [4, pp.491–493, Definition A.12 and Definition A.18].

# 5  Axiomatization and Decidability of LHS⁻

In this section, we turn our attention to LHS⁻. Precisely, we will provide a proof system for the logic, which is also helpful to show that its satisfiability problem is decidable. To achieve the former, instead of applying directly the usual techniques involving canonical models, we will make a detour: very roughly, we will first separate the 'black part' and the 'white part' of the language $\mathcal{L}^-$ of LHS⁻ and then build a desired calculus on that for the standard modal logic K. The details will indicate that containing two kinds of propositional variables in $\mathcal{L}^-$ makes LHS⁻ very different from its counterpart K × K in product logic. Let us now introduce the details.

A formula $\varphi \in \mathcal{L}^-$ is *clean* if it contains only black modal operators or white modal operators. Formulas in the language $\mathcal{L}^-$ of LHS⁻ may contain nested black modalities and white modalities. However, as we shall see, *every $\varphi \in \mathcal{L}^-$ is logically equivalent to a Boolean combination of some clean formulas.*

**Definition 6.** *Languages $\mathcal{L}_\square$ and $\mathcal{L}_\blacksquare$ are given by:*

$$\mathcal{L}_\square \ni \varphi ::= p^l \mid \neg\varphi \mid \varphi \wedge \varphi \mid \square\varphi,$$
$$\mathcal{L}_\blacksquare \ni \varphi ::= p^r \mid \neg\varphi \mid \varphi \wedge \varphi \mid \blacksquare\varphi,$$

*where $p^l \in \mathsf{L}$ and $p^r \in \mathsf{R}$.*

Let $\mathsf{K}_\square$ and $\mathsf{K}_\blacksquare$ denote *the minimal modal logics with the languages $\mathcal{L}_\square$ and $\mathcal{L}_\blacksquare$*, respectively. As the case for the standard modal logic K, the satisfiability problems for both the logics are decidable (cf. [1]). Also, except the difference of the languages, their proof systems are the same as that for K [4], and we write $\mathbf{K}_\square$ and $\mathbf{K}_\blacksquare$ for them. In what follows, we write $\models_1$ for *the usual one-dimensional satisfaction relation*. By induction on formulas, we can show that:

**Proposition 7.** *Let $\mathfrak{M} = (W, R, V)$ be a $\mathsf{K}_\square$-model and $\mathfrak{N} = (U, S, V')$ a $\mathsf{K}_\blacksquare$-model such that $W \cap U = \emptyset$. Let $\psi \in \mathcal{L}_\square$, $\gamma \in \mathcal{L}_\blacksquare$. Then for all $s \in W$ and $t \in U$,*

*(1) $\mathfrak{M}, s \models_1 \psi$ if and only if $\mathfrak{M} \uplus \mathfrak{N}, s, t \models \psi$, and*
*(2) $\mathfrak{N}, t \models_1 \gamma$ if and only if $\mathfrak{M} \uplus \mathfrak{N}, s, t \models \gamma$,*

*where $\mathfrak{M} \uplus \mathfrak{N}$ is the LHS-model defined by $\mathfrak{M} \uplus \mathfrak{N} = (W \cup U, R \cup S, V \cup V')$. The LHS-model $\mathfrak{M} \uplus \mathfrak{N}$ is called the disjoint union of $\mathfrak{M}$ and $\mathfrak{N}$.*

Let $\mathfrak{M}$ be a $\mathsf{K}_\square$-model and $\mathfrak{N}$ a $\mathsf{K}_\blacksquare$-model. Since there are always isomorphic copies of them with disjoint domains, we can always assume that the domains of $\mathfrak{M}$ and $\mathfrak{N}$ are disjoint and construct the disjoint union of $\mathfrak{M}$ and $\mathfrak{N}$.

For an arbitrary LHS-model $\mathfrak{M} = (W, R, V)$, by restricting the valuation to $\mathsf{L}$ (we write $V|_\mathsf{L}$ for it), we can obtain a model $\mathfrak{M}|_\mathsf{L} = (W, R, V|_\mathsf{L})$ for $\mathcal{L}_\square$, and similarly, by restricting $V$ to $\mathsf{R}$ (we write $V|_\mathsf{R}$ for it), we can get a model $\mathfrak{M}|_\mathsf{R} = (W, R, V|_\mathsf{R})$ for $\mathcal{L}_\blacksquare$. By induction on formulas, it is simple to prove that

**Proposition 8.** *Let $\langle \mathfrak{M}, s, s \rangle$ be a pointed LHS-model. Then, for any $\varphi \in \mathcal{L}_\square$, $\mathfrak{M}, s, s \models \varphi$ iff $\mathfrak{M}|_\mathsf{L}, s \models_1 \varphi$. Also, for any $\varphi \in \mathcal{L}_\blacksquare$, $\mathfrak{M}, s, s \models \varphi$ iff $\mathfrak{M}|_\mathsf{R}, s \models_1 \varphi$.*

Before the next step, let us first recall some concepts and facts about propositional logic. Let $\mathcal{L}_p$ denote the propositional language whose propositional variables come from $\mathsf{L} \cup \mathsf{R}$. For each $\varphi \in \mathcal{L}_p$, we write $\varphi(p_1, \ldots, p_n)$ if the propositional variables occurring in $\varphi$ are among $p_1, \ldots, p_n$. Let $\varphi(\alpha_1, \ldots, \alpha_n)$ denote the formula obtained from $\varphi(p_1, \ldots, p_n)$ by simultaneously substituting $p_1, \ldots, p_n$ with $\alpha_1, \ldots, \alpha_n$ respectively. Let PL denote the set of all valid formulas in $\mathcal{L}_p$. A sound and complete Hilbert style calculus **PL** for PL can be given in a usual way.

**Definition 7.** *A formula $\varphi \in \mathcal{L}_p$ is in conjunctive normal form (CNF), if $\varphi$ is of the form $\bigwedge_{i=1}^{n} \bigvee_{j=1}^{m_i} \varphi_{ij}$, where $n, m_1, \ldots, m_n \in \mathbb{N}^+$ and each $\varphi_{ij}$ is a propositional variable or a negation of a propositional variable.*

We say that a formula $\varphi$ is a *CNF-formula* if $\varphi$ is in conjunctive normal form. Let $\mathsf{CNF}_p$ denote the set of all formulas $\varphi \in \mathcal{L}_p$ in CNF.

**Proposition 9.** *There is a function $h : \mathcal{L}_p \to \mathsf{CNF}_p$ such that for all $\varphi \in \mathcal{L}_p$, $\vdash_{\mathbf{PL}} \varphi \leftrightarrow h(\varphi)$.*

*Proof.* Such a function can be found in many textbooks of mathematical logic (see e.g., [6, p. 221, Theorem 4.7]). □

**Definition 8.** *Let $\varphi \in \mathcal{L}^-$. Then we say that $\varphi \in \mathcal{L}^-$ is a clean formula, if there are $\psi_1, \ldots, \psi_n \in \mathcal{L}_\square$, $\gamma_1, \ldots, \gamma_m \in \mathcal{L}_\blacksquare$ and $\alpha(p_1^l, \ldots, p_n^l, p_1^r, \ldots, p_m^r) \in \mathcal{L}_p$ such that $\varphi = \alpha(\psi_1, \ldots, \psi_n, \gamma_1, \ldots, \gamma_m)$. Moreover, if $\alpha$ is in CNF, then $\varphi$ is called a clean CNF-formula. Let $\mathcal{L}_c$ and $\mathsf{CNF}_c$ denote the set of all clean formulas and the set of all clean CNF-formulas, respectively.*

Table 1 presents a Hilbert style calculus **LHS$^-$** for LHS$^-$, which is a direct extension of the calculi **PL**, **K$_\square$** and **K$_\blacksquare$**. Therefore, we have the following:

**Table 1.** A proof system **LHS$^-$** for LHS$^-$

| Proof system **LHS$^-$** for LHS$^-$ |
| --- |
| **Axiom schemes:** |
| (A1)    $\alpha \to (\beta \to \alpha)$ |
| (A2)    $(\alpha \to (\beta \to \theta)) \to ((\alpha \to \beta) \to (\alpha \to \theta))$ |
| (A3)    $(\neg\beta \to \neg\alpha) \to (\alpha \to \beta)$ |
| (K)    $\boxtimes(\alpha \to \beta) \to (\boxtimes\alpha \to \boxtimes\beta)$, for $\boxtimes \in \{\square, \blacksquare\}$ |
| (R$_\square$)    $\square(\psi \vee \gamma) \leftrightarrow (\square\psi \vee \gamma)$, where $\psi \in \mathcal{L}_\square$ and $\gamma \in \mathcal{L}_\blacksquare$ |
| (R$_\blacksquare$)    $\blacksquare(\psi \vee \gamma) \leftrightarrow (\psi \vee \blacksquare\gamma)$, where $\psi \in \mathcal{L}_\square$ and $\gamma \in \mathcal{L}_\blacksquare$ |
| **Inference rules:** |
| (MP)    From $\alpha$ and $\alpha \to \beta$, infer $\beta$ |
| (Nec$_\boxtimes$)    From $\alpha$, infer $\boxtimes\alpha$, for $\boxtimes \in \{\square, \blacksquare\}$ |

**Proposition 10.** *For all $\varphi \in \mathcal{L}_p$, if $\vdash_{\mathbf{PL}} \varphi$, then $\vdash_{\mathbf{LHS}^-} \varphi$.*

**Proposition 11.** *For any formula $\varphi$ of $\mathcal{L}_\square$, if $\vdash_{\mathbf{K}_\square} \varphi$, then $\vdash_{\mathbf{LHS}^-} \varphi$. Similarly, for any formula $\varphi$ of $\mathcal{L}_\blacksquare$, if $\vdash_{\mathbf{K}_\blacksquare} \varphi$, then $\vdash_{\mathbf{LHS}^-} \varphi$.*

**Corollary 2.** *There is a function $f : \mathcal{L}_c \to \mathsf{CNF}_c$ such that for all $\varphi \in \mathcal{L}_c$, it holds that $\vdash_{\mathbf{LHS}^-} \varphi \leftrightarrow f(\varphi)$. Also, the resulting formula $f(\varphi)$ is of the form $\bigwedge_{i=1}^n (\psi_i \vee \gamma_i)$, where $\bigwedge_{i=1}^n \psi_i \in \mathcal{L}_\square$ and $\bigwedge_{i=1}^n \gamma_i \in \mathcal{L}_\blacksquare$.*

*Proof.* Let $\varphi$ be a clean formula. Then, there are formulas $\beta(p_1, \ldots, p_k) \in \mathcal{L}_p$ and $\alpha_1, \ldots, \alpha_k \in \mathcal{L}_\square \cup \mathcal{L}_\blacksquare$ such that $\varphi = \beta(\alpha_1, \ldots, \alpha_k)$. It follows from Proposition 9 that $\vdash_{\mathbf{PL}} \beta \leftrightarrow h(\beta)$. Note that $h(\beta)$ is in CNF, and so $h(\beta)$ is of the form $\bigwedge_{i=1}^n \bigvee_{j=1}^{m_i} \beta_{ij}$ with $n, m_1, \ldots, m_n \in \mathbb{N}^+$. For each $1 \leq i \leq n$, we define:

- $\psi_i := (p^l \wedge \neg p^l) \vee \bigvee \{\beta_{ij} \in \mathcal{L}_\square : 1 \leq j \leq m_i\}$, and
- $\gamma_i := (p^r \wedge \neg p^r) \vee \bigvee \{\beta_{ij} \in \mathcal{L}_\blacksquare : 1 \leq j \leq m_i\}$,

where $p^l \in \mathsf{L}$ and $p^r \in \mathsf{R}$ are new propositional variables. Then it holds that

$$\vdash_{\mathbf{PL}} \bigwedge_{i=1}^n \bigvee_{j=1}^{m_i} \beta_{ij} \leftrightarrow \bigwedge_{i=1}^n (\psi_i \vee \gamma_i).$$

By Proposition 10, $\vdash_{\mathbf{LHS}^-} \beta(p_1, \ldots, p_k) \leftrightarrow \bigwedge_{i=1}^n (\psi_i \vee \gamma_i)$. Next, applying the inference rule (Sub) of the calculus $\mathbf{LHS}^-$, we can obtain

$$\vdash_{\mathbf{LHS}^-} \varphi \leftrightarrow \bigwedge_{i=1}^n (\psi_i(\alpha_1, \ldots, \alpha_k, p^l) \vee \gamma_i(\alpha_1, \ldots, \alpha_k, p^r)).$$

Now, we can define a desired function $f : \mathcal{L}_c \to \mathsf{CNF}_c$ in the following manner:

$$f(\varphi) = \bigwedge_{i=1}^n (\psi_i(\alpha_1, \ldots, \alpha_k, p^l) \vee \gamma_i(\alpha_1, \ldots, \alpha_k, p^r)),$$

which completes the proof. □

**Lemma 3.** *Let $\mathfrak{M} = (W, R, V)$ be a model and $s, t \in W$. Then, for all formulas $\varphi \in \mathcal{L}_\square$ and $\psi \in \mathcal{L}_\blacksquare$, the following equivalences hold:*

*(1) $\mathfrak{M}, s, t \models \varphi$ if, and only if, for all $t' \in W$, $\mathfrak{M}, s, t' \models \varphi$.*
*(2) $\mathfrak{M}, s, t \models \psi$ if, and only if, for all $s' \in W$, $\mathfrak{M}, s', t \models \varphi$.*
*(3) $\mathfrak{M}, s, t \models \square(\varphi \vee \psi)$ if, and only if, $\mathfrak{M}, s, t \models \square\varphi \vee \psi$.*
*(4) $\mathfrak{M}, s, t \models \blacksquare(\varphi \vee \psi)$ if, and only if, $\mathfrak{M}, s, t \models \varphi \vee \blacksquare\psi$.*

*Proof.* The proofs for items (1) and (2) are by induction on the complexity of $\varphi$ and $\psi$, respectively. We omit the details for them. In what follows, we merely consider for (3), since (4) can be proved in a similar way.

For the direction from left to right, we prove the contrapositive and assume that $\mathfrak{M}, s, t \not\models \square\varphi \vee \psi$. Then, $\mathfrak{M}, s, t \models \neg\psi$ and there is some state $s' \in R(s)$ such that $\mathfrak{M}, s', t \not\models \varphi$. Note that $\neg\psi \in \mathcal{L}_\blacksquare$. Now, using the item (2), we can obtain $\mathfrak{M}, s', t \models \neg\psi$. Thus, it holds that $\mathfrak{M}, s', t \not\models \varphi \vee \psi$, and so $\mathfrak{M}, s, t \not\models \square(\varphi \vee \psi)$.

For the converse direction, we assume that $\mathfrak{M}, s, t \not\models \Box(\varphi \vee \psi)$. Then, there is some state $s' \in R(s)$ s.t. $\mathfrak{M}, s', t \not\models \varphi \vee \psi$, which entails $\mathfrak{M}, s, t \not\models \Box\varphi$ and $\mathfrak{M}, s', t \not\models \psi$. Using item (2), we can infer $\mathfrak{M}, s, t \not\models \psi$ from $\mathfrak{M}, s', t \not\models \psi$. Hence, $\mathfrak{M}, s, t \not\models \Box\varphi \vee \psi$. The proof is completed. □

From the items (3) and (4) of Lemma 3, it is a matter of direct checking that:

**Lemma 4.** *For all $\varphi \in \mathcal{L}_\Box$ and $\psi \in \mathcal{L}_\blacksquare$, both the formulas $\Box(\varphi \vee \psi) \leftrightarrow (\Box\varphi \vee \psi)$ and $\blacksquare(\varphi \vee \psi) \leftrightarrow (\varphi \vee \blacksquare\psi)$ are valid.*

Now we move to showing the soundness of the proof system **LHS⁻**:

**Theorem 3.** *For any formula $\varphi \in \mathcal{L}^-$, $\vdash_{\mathbf{LHS}^-} \varphi$ implies $\models \varphi$.*

*Proof.* The validity of (A1), (A2), (A3), (K$_\Box$) and (K$_\blacksquare$) is easy to see. Also, Lemma 4 indicates that (R$_\Box$) and (R$_\blacksquare$) are valid. Moreover, all inference rules of **LHS⁻** preserve validity, and the details are left as an exercise. □

Next, we consider for the completeness of the calculus **LHS⁻**.

**Definition 9.** *For any $\mathcal{L}^-$-formula $\varphi$, we define its clean CNF companion $\varphi_c$ in the following inductive manner:*

$$(p^l)_c := p^l \vee (p_0^r \wedge \neg p_0^r), \text{ where } p_0^r \text{ is a new propositional variable from } \mathsf{R}.$$
$$(p^r)_c := (p_0^l \wedge \neg p_0^l) \vee p^r, \text{ where } p_0^l \text{ is a new propositional variable from } \mathsf{L}.$$
$$(\neg\varphi)_c := f(\neg\varphi_c)$$
$$(\varphi \wedge \psi)_c := \varphi_c \wedge \psi_c$$
$$(\Box\varphi)_c := \bigwedge_{i=1}^n (\Box\psi_i \vee \gamma_i), \text{ where } \bigwedge_{i=1}^n \psi_i \in \mathcal{L}_\Box, \bigwedge_{i=1}^n \gamma_i \in \mathcal{L}_\blacksquare \text{ and } \varphi_c = \bigwedge_{i=1}^n (\psi_i \vee \gamma_i).$$
$$(\blacksquare\varphi)_c := \bigwedge_{i=1}^n (\psi_i \vee \blacksquare\gamma_i), \text{ where } \bigwedge_{i=1}^n \psi_i \in \mathcal{L}_\Box, \bigwedge_{i=1}^n \gamma_i \in \mathcal{L}_\blacksquare \text{ and } \varphi_c = \bigwedge_{i=1}^n (\psi_i \vee \gamma_i).$$

*Example 1.* Let us consider an example $\blacksquare p^l$. With the clauses above, we have $(\blacksquare p^l)_c = p^l \vee \blacksquare(p_0^r \wedge \neg p_0^r)$. Note that if we define $(p^l)_c$ to be $p^l$, then we cannot ensure that a formula and its clean companion are equivalent: for instance, one can easily find a model making $\blacksquare p^l \leftrightarrow p^l$ false. Similarly for the clause of $p^r \in \mathsf{R}$.

**Theorem 4.** *Let $\varphi$ be a formula of $\mathcal{L}^-$. Then, its clean CNF companion $\varphi_c$ is of the form $\bigwedge_{i=1}^n (\psi_i \vee \gamma_i)$ with $\bigwedge_{i=1}^n \psi_i \in \mathcal{L}_\Box$ and $\bigwedge_{i=1}^n \gamma_i \in \mathcal{L}_\blacksquare$. Moreover, it holds that $\vdash_{\mathbf{LHS}^-} \varphi \leftrightarrow \varphi_c$.*

*Proof.* The first part of the theorem is easy, since $\varphi_c$ is a clean CNF formula. In what follow, by induction on $\varphi \in \mathcal{L}^-$, we will show that $\vdash_{\mathbf{LHS}^-} \varphi \leftrightarrow \varphi_c$. The cases for propositional atoms and $\wedge$ are straightforward, and we consider others.

(1). First, we consider $\varphi = \neg\psi$. By the induction hypothesis, it holds that $\vdash_{\mathbf{LHS}^-} \psi \leftrightarrow \psi_c$. So, $\vdash_{\mathbf{LHS}^-} \neg\psi \leftrightarrow \neg\psi_c$. Clearly, $\neg\psi_c \in \mathcal{L}_c$. From Corollary 2 we know that $\vdash_{\textbf{LHS}^-} f(\neg\psi_c) \leftrightarrow \neg\psi_c$. Also, with the clause for $\neg$ in Definition 9, we have $(\varphi)_c = f(\neg\psi_c)$. Immediately, we obtain $\vdash_{\textbf{LHS}^-} \varphi_c \leftrightarrow \varphi$, as desired.

(2). Next, we consider $\varphi = \Box\psi$. Assume that $\psi_c = \bigwedge_{i=1}^{n}(\psi_i' \vee \gamma_i')$, where $\bigwedge_{i=1}^{n}\psi_i' \in \mathcal{L}_\Box$ and $\bigwedge_{i=1}^{n}\gamma_i' \in \mathcal{L}_\blacksquare$. Then, $\vdash_{\textbf{LHS}^-} \Box\psi_c \leftrightarrow \bigwedge_{i=1}^{n}\Box(\psi_i' \vee \gamma_i')$. For simplicity, we write $\gamma(p^l, p^r)$ for $\Box(p^l \vee p^r) \leftrightarrow (\Box p^l \vee p^r)$, which is exactly the axiom $(R_\Box)$. Note that for each $1 \leq i \leq n$, $\psi_i' \in \mathcal{L}_\Box$ and $\gamma_i' \in \mathcal{L}_\blacksquare$. So, for each $1 \leq i \leq n$, using the inference rule (Sub), we can obtain $\vdash_{\textbf{LHS}^-} \gamma(\psi_i', \gamma_i')$, i.e., $\vdash_{\textbf{LHS}^-} \Box(\psi_i' \vee \gamma_i') \leftrightarrow (\Box\psi_i' \vee \gamma_i')$. Therefore, $\vdash_{\textbf{LHS}^-} \bigwedge_{i=1}^{n}\Box(\psi_i' \vee \gamma_i') \leftrightarrow \bigwedge_{i=1}^{n}(\Box\psi_i' \vee \gamma_i')$, which entails $\vdash_{\textbf{LHS}^-} \Box\psi_c \leftrightarrow \varphi_c$. By induction hypothesis, $\vdash_{\textbf{LHS}^-} \psi \leftrightarrow \psi_c$ and so $\vdash_{\textbf{LHS}^-} \varphi \leftrightarrow \Box\psi_c$. Hence $\vdash_{\textbf{LHS}^-} \varphi \leftrightarrow \varphi_c$.

(3). Finally, the case for $\varphi = \blacksquare\psi$ is similar to (2). The proof is completed. □

*Example 2.* Applications of Theorem 4 can be diverse. As an example, we show how to use it to prove $\vdash_{\textbf{LHS}^-} \Box\blacksquare\neg\varphi \leftrightarrow \blacksquare\Box\neg\varphi$. Let $(\neg\varphi)_c = \bigwedge_{i=1}^{n}(\psi_i \vee \gamma_i)$ with $\bigwedge_{i=1}^{n}\psi_i \in \mathcal{L}_\Box$ and $\bigwedge_{i=1}^{n}\gamma_i \in \mathcal{L}_\blacksquare$. By Theorem 4, $\vdash_{\textbf{LHS}^-} \neg\varphi \leftrightarrow (\neg\varphi)_c$. Then, $\vdash_{\textbf{LHS}^-} \blacksquare\Box\neg\varphi \leftrightarrow \blacksquare\Box(\neg\varphi)_c$ and $\vdash_{\textbf{LHS}^-} \Box\blacksquare\neg\varphi \leftrightarrow \Box\blacksquare(\neg\varphi)_c$. Also, we have $\vdash_{\textbf{LHS}^-} \blacksquare\Box(\neg\varphi)_c \leftrightarrow \bigwedge_{i=1}^{n}(\Box\psi_i \vee \blacksquare\gamma_i)$ and $\vdash_{\textbf{LHS}^-} \Box\blacksquare(\neg\varphi)_c \leftrightarrow \bigwedge_{i=1}^{n}(\Box\psi_i \vee \blacksquare\gamma_i)$. Thus, we obtain $\vdash_{\textbf{LHS}^-} \Box\blacksquare\neg\varphi \leftrightarrow \blacksquare\Box\neg\varphi$.

Now, with the help of Theorem 4, we can show that the proof system $\textbf{LHS}^-$ is complete with respect to the class $\textsf{Mod}^{<\omega}$ of finite models:

**Theorem 5.** *For each $\varphi \in \mathcal{L}^-$, if $\textsf{Mod}^{<\omega} \models \varphi$ then $\vdash_{\textbf{LHS}^-} \varphi$.*

*Proof.* Let $\varphi \in \mathcal{L}^-$ and $\nvdash_{\textbf{LHS}^-} \varphi$. By Theorem 4, $\nvdash_{\textbf{LHS}^-} \varphi_c$ and $\varphi_c$ is of the form $\bigwedge_{i=1}^{n}(\psi_i \vee \gamma_i)$ where $\bigwedge_{i=1}^{n}\psi_i \in \mathcal{L}_\Box$ and $\bigwedge_{i=1}^{n}\gamma_i \in \mathcal{L}_\blacksquare$. Since $\nvdash_{\textbf{LHS}^-} \varphi_c$, there is some $i$ such that $\nvdash_{\textbf{LHS}^-} \psi_i \vee \gamma_i$. By Proposition 11, we have $\nvdash_{\textbf{K}_\Box} \psi_i$ and $\nvdash_{\textbf{K}_\blacksquare} \gamma_i$. By the completeness of $\textbf{K}_\Box$ and $\textbf{K}_\blacksquare$, both $\neg\psi_i$ and $\neg\gamma_i$ is satisfiable. Since $\textbf{K}_\Box$ and $\textbf{K}_\blacksquare$ have the finite model property [4], there are finite pointed $\textbf{K}_\Box$-model $\langle \mathfrak{M}, s \rangle$ and finite pointed $\textbf{K}_\blacksquare$-model $\langle \mathfrak{N}, t \rangle$ such that $\mathfrak{M}, s \nvDash \psi_i$ and $\mathfrak{N}, t \nvDash \gamma_i$. Then, by Proposition 7, $\mathfrak{M} \uplus \mathfrak{N}, s, t \models \neg\psi_i \wedge \neg\gamma_i$, which entails $\mathfrak{M} \uplus \mathfrak{N}, s, t \models \neg\varphi_c$. By Theorem 3 and Theorem 4, $\textsf{Mod}^{<\omega} \nvDash \varphi$. □

**Theorem 6.** $\textsf{LHS}^-$ *enjoys the finite model property, and it is decidable.*

*Proof.* Assume that $\varphi \in \mathcal{L}^-$ is satisfiable. By the soundness of $\textbf{LHS}^-$ (Theorem 3), we have $\nvdash_{\textbf{LHS}^-} \neg\varphi$. Then, it follows from Theorem 5 that there is some finite model $\mathfrak{M}'$ satisfying $\varphi$. So, the first claim holds. Now, since $\textsf{LHS}^-$ can be finitely axiomatized and has the finite model property, the logic is decidable. □

## 6   Conclusion

*Summary* The article is a technical continuation of [15,16], which explored the properties LHS, a tool to reason about the games of hide and seek. In the paper, we show that the satisfiability problem for LHS with a single relation is undecidable. Also, based on existing notions of bisimulations and first-order translation for LHS, we provide a van Benthem style characterization theorem for the logic. Moreover, we develop a Hilbert style calculus for $\mathsf{LHS}^-$ and prove that its satisfiability problem is decidable, and the details of our proofs are helpful to understand the differences between $\mathsf{LHS}^-$ and $\mathsf{K} \times \mathsf{K}$. All these results can be transferred to logics generalizing $\mathsf{LHS}$ and $\mathsf{LHS}^-$ for games with $n > 2$ players.

*Related Work* As stated, our work is closely related to product logics $\mathsf{K} \times \mathsf{K}$ [8] and $\mathsf{K} \times^\delta \mathsf{K}$ [9,11,12], cylindric modal logics [21] and cylindric algebra [10]. Also, there is a line of logical investigation for the hide and seek game. Needless to say, the most relevant ones are [15] and its extension [16]. The latter offers a notion of bisimulation for $\mathsf{LHS}$ and its first-order translation, proves the undecidability of $\mathsf{LHS}$ with multiple relations, shows that the model-checking problems for both $\mathsf{LHS}$ and $\mathsf{LHS}^-$ are P-complete, and identifies the counterpart of $\mathsf{LHS}$ in product logic. Moreover, [19] extends $\mathsf{LHS}$ and $\mathsf{LHS}^-$ with components from hybrid logics, and study the expressiveness and axiomatization of the resulting logics. Also, [14] develops logical tools to capture how players update their knowledge about other players' positions in the hide and seek game, in which players have only imperfect information. Finally, it is important to notice that besides the efforts made for the hide and seek game, many other graph games and their matching modal logics have also been studied in recent years, and we refer to [2] for a broad research program on this topic and refer to [13] for extensive references to modal logics for graph games.

*Further Directions* Except what we have explored in the article, there are a number of directions deserving to be explored in future. A natural next step is to explore the exact complexity of $\mathsf{LHS}$ and $\mathsf{LHS}^-$. Also, it is important to study the axiomatizability of $\mathsf{LHS}$, for which it might be useful to analyze the techniques developed for $\mathsf{K} \times^\delta \mathsf{K}$ [11]. Closely related to this, [15,16] provide preliminary discussions on the difference between the frameworks of $\mathsf{LHS}$ and $\mathsf{LHS}^-$ and product logics, but their exact relation remains to be explored. For expressiveness, besides the expressive power of $\mathsf{LHS}$ w.r.t. models, the equality constant $I$ improves the frame definability of $\mathsf{LHS}$ as well,[5] and it remains to have a comprehensive understanding of the expressive power of $\mathsf{LHS}$ w.r.t. frames. Moreover, another direction is to develop the proof theory for our logics, and for instance, provide sequent calculi and tableau systems for them. Finally, it is worthwhile to generalize our results to broader settings and consider the extensions of $\mathsf{LHS}$ and $\mathsf{LHS}^-$ with further operators, such as *graded modalities* [7,20] to talk about the degree of a state in a graph.

**Acknowledgements.** We thank Johan van Benthem, Sujata Ghosh, Fenrong Liu and Katsuhiko Sano for their inspiring suggestions, and thank the three anonymous reviewers for their helpful comments for improvements. Qian Chen is supported by Tsinghua University Initiative Scientific Research Program. Dazhu Li is supported by National Social Science Foundation of China [22CZX063].

---

[5] For instance, although *confluence* (i.e., $\forall x \forall y_1 \forall y_2 (Rxy_1 \wedge Rxy_2 \rightarrow \exists z(Ry_1 z \wedge Ry_2 z)))$ is not definable with the basic modal language, with the techniques of *frame correspondence* [3], one can check that the property can be simply defined as $I \rightarrow \square\blacksquare\lozenge\blacklozenge I$.

# References

1. van Benthem, J.: Modal Logic for Open Minds. CSLI Publications (2010)
2. van Benthem, J., Liu, F.: Graph games and logic design. In: Liu, F., Ono, H., Yu, J. (eds.) Knowledge, Proof and Dynamics. LASLL, pp. 125–146. Springer, Singapore (2020). https://doi.org/10.1007/978-981-15-2221-5_7
3. van Benthem, J.: Correspondence theory. In: Gabbay, D., Guenthner, F. (eds.) Handbook of Philosophical Logic, Synthese Library (Studies in Epistemology, Logic, Methodology, and Philosophy of Science), vol. 165, pp. 167–247. Springer, Dordrecht (1984). https://doi.org/10.1007/978-94-009-6259-0_4
4. Blackburn, P., de Rijke, M., Venema, Y.: Modal Logic. Cambridge University Press, Cambridge (2001)
5. Chang, C., Keisler, H.: Model Theory, Studies in Logic and the Foundations of Mathematics, 2nd edn., vol. 73. Elsevier, Amsterdam (1990)
6. Ebbinghaus, H.D., Flum, J., Thomas, W.: Mathematical Logic, 3rd edn. Springer, Cham (2021). https://doi.org/10.1007/978-3-030-73839-6
7. Fine, K.: In so many possible worlds. Notre Dame J. Form. Log. **13**, 516–520 (1972)
8. Gabbay, D., Kurucz, A., Wolter, F., Zakharyaschev, M.: Many-Dimensional Modal Logics: Theory and Applications. Elsevier, Amsterdam (2003)
9. Hampson, C., Kikot, S., Kurucz, A.: The decision problem of modal product logics with a diagonal, and faulty counter machines. Stud. Logica. **104**, 455–486 (2016)
10. Henkin, L., Monk, J.D., Tarski, A.: Cylindric Algebras, Part I, Studies in Logic and the Foundations of Mathematics, vol. 64. North-Holland, Amsterdam (1971)
11. Kikot, S.: Axiomatization of modal logic squares with distinguished diagonal. Math. Notes **88**, 238–250 (2010)
12. Kurucz, A.: Products of modal logics with diagonal constant lacking the finite model property. In: Ghilardi, S., Sebastiani, R. (eds.) FroCoS 2009. LNCS (LNAI), vol. 5749, pp. 279–286. Springer, Heidelberg (2009). https://doi.org/10.1007/978-3-642-04222-5_17
13. Li, D.: Formal Threads in the Social Fabric: Studies in the Logical Dynamics of Multi-Agent Interaction. Ph.D. thesis, Department of Philosophy, Tsinghua University and ILLC, University of Amsterdam (2021)
14. Li, D., Ghosh, S., Liu, F.: Dynamic-epistemic logic for the cops and robber game. Manuscript (2023)
15. Li, D., Ghosh, S., Liu, F., Tu, Y.: On the subtle nature of a simple logic of the hide and seek game. In: Silva, A., Wassermann, R., de Queiroz, R. (eds.) WoLLIC 2021. LNCS, vol. 13038, pp. 201–218. Springer, Cham (2021). https://doi.org/10.1007/978-3-030-88853-4_13
16. Li, D., Ghosh, S., Liu, F., Tu, Y.: A simple logic of the hide and seek game. Stud. Logica. **111**, 821–853 (2023). https://doi.org/10.1007/s11225-023-10039-4
17. Nowakowski, R., Winkler, R.P.: Vertex-to-vertex pursuit in a graph. Discrete Math. **43**, 235–239 (1983)
18. Robinson, R.: Undecidability and nonperiodicity for tilings of the plane. Invent. Math. **12**, 177–209 (1971). https://doi.org/10.1007/BF01418780
19. Sano, K., Liu, F., Li, D.: Hybrid logic of the hide and seek game. Manuscript (2023)
20. Sano, K., Ma, M.: Goldblatt-Thomason-style theorems for graded modal language. In: Beklemishev, L., Goranko, V., Shehtman, V. (eds.) Advances in Modal Logic, vol. 8, pp. 330–349. CSLI Publications, San Diego (2010)
21. Venema, Y.: Many-Dimensional Modal Logics. Ph.D. thesis, Universiteit van Amsterdam (1991)

# Axiomatization of Hybrid Logic of Link Variations

Penghao Du[1] and Qian Chen[1,2(✉)] ⓘ

[1] The Tsinghua-UvA Joint Research Centre for Logic, Department of Philosophy, Tsinghua University, Beijing, China
{dph21,chenq21}@mails.tsinghua.edu.cn
[2] Institute for Logic, Language and Computation, University of Amsterdam, Amsterdam, The Netherlands

**Abstract.** In this paper, we investigate local and global dynamic modal operators which have the ability to update the accessibility relation of a model. For the global operators, the logic GLV of global link variations based on hybrid logic H(@) is introduced, which involves global link cutting, adding and rotating simultaneously. A Hilbert-style calculus $C_{GLV}$ is provided. By constructing families of canonical models inductively, we prove that the calculus $C_{GLV}$ is sound and strongly complete with respect to GLV. For the local operators, we extended the logic LLD(@, ↓) of link deletion introduced in [12] to LLV, which is based on the hybrid logic H(E, ↓) and involves local definable link cutting, adding and rotating. By defining local named dynamic operators and providing recursion axioms for them, we obtain a sound and strongly complete calculus $C_{LLV}$ for LLV. Moreover, we show that for an arbitrary set $X$ of global and local dynamic operators, the calculus $C_{GLV}(X)$ and $C_{LLV}(X)$ are still sound and strongly complete w.r.t the logic GLV($X$) and LLV($X$), respectively.

**Keywords:** Dynamic logic · Hybrid logic · Axiomatization

## 1 Introduction

Link variations in graph theory have been widely studied in recent years, with applications in many areas such as knowledge graph, social network, and graph game (cf. [7,11,19]). Link cutting and adding are crucial operations in link variation. These operations play important role in the *knowledge graph*, which is a critical area in AI (cf. [11]). For example, the update of *heterogeneous graph* in Fig. 1 involves both link cutting and adding simultaneously.

Graph games are widely investigated by logicians and many results on game logics have been obtained in recent years (cf. [1,3,12]). Dynamic graph game is a special kind of graph game during which the edges of graphs change. Link variations play a central role in these kind of games. *Sabotage game* is a classical example of game logic, in which global link cutting is the key dynamic operation (cf. [3,6,15]). In the sabotage game, there are two players and a directed graph

N. Gierasimczuk and F. R. Velázquez-Quesada (Eds.): DaLí 2023, LNCS 14401, pp. 35–51, 2024.
https://doi.org/10.1007/978-3-031-51777-8_3

with a winning area. One player (the *traveler*) aims to move successfully to the winning area, while the other player (the *demon*) tries to prevent the traveler from reaching her destinations by globally cutting one link in each round. Sabotage modal logic SML was introduced in [13]. It is proved in [14] that SML over edge-labelled transition systems is undecidable and lacks the finite model property. In [2], it is proved that the logic SML is undecidable. Some decidable fragments of SML are introduced in [2] by giving translations from global relation-changing modal logics to hybrid logic with downarrow. Moreover, an axiomatization of hybrid sabotage modal logic was presented in [8].

**Fig. 1.** Updates on a transportation network

Influenced by sabotage game, [12] introduced the sabotage game with local link deletion, which is the same as sabotage game except that demon locally cut a branch of links in each round. As it is claimed in [1], local link cutting operations are essentially different from the global one, which makes these two games and their modal logics quite different. In [1], the logic SML is extended by link adding and rotating operators. Expressive power and model checking problem of these logics are studied.

As it is shown above, both the sabotage game and its logic are widely studied and there are many games and logics generalized from them, respectively. However, there is no systematic analysis of axiomatization of these dynamic logics, especially for those with different kinds of link variations operators. In this work, based on hybrid logic, we first extend SML to GLV with global link adding and rotating operators. A sound and complete axiomatization for GLV is provided. Moreover, we introduce the logic LLV of local definable link variations and provide also an axiomatization for it.

The paper is structured as follows. Section 2 gives preliminaries of the logic GLV of global link variations, which is based on the hybrid logic H(@) and involves link cutting, adding and rotating simultaneously. Section 3 provides a Hilbert-style calculus $C_{GLV}$ for GLV and shows the soundness of $C_{GLV}$. In Sect. 4, we prove that for an arbitrarily chosen subset $X$ of dynamic operators of $\{+, -, \circlearrowright\}$, the calculus $C_{GLV}(X)$ is strongly complete with respect to GLV($X$). In Sect. 5, we breifly discuss the logic LLV of local (definable) link variations. Axiomatizations of these logics are provided, and we show their strongly completeness by recursion axioms and local named operators.

## 2     Logic of Global Link Variation

### 2.1     Preliminaries of **GLV**

We start by introducing the formal language of the logic GLV, which is a modal logic based on the hybrid logic H(@) (cf. [10]). Let Prop $= \{p_n : n \in \omega\}$ be a countable set of propositional variables.

**Definition 1.** *Let* Nom $= \{a_n : n \in \omega\}$ *be a set of nominals which is disjoint from* Prop. *The language* $\mathcal{L}_{\mathsf{Nom}}$ *of* GLV *over* Nom *is defined as follows:*

$$\mathcal{L}_{\mathsf{Nom}} \ni \varphi ::= a \mid p \mid \neg\varphi \mid \varphi \wedge \psi \mid \Diamond\varphi \mid @_a\varphi \mid \langle + \rangle\varphi \mid \langle - \rangle\varphi \mid \langle \circlearrowleft \rangle\varphi$$

*where* $p \in$ Prop *and* $a \in$ Nom. *Abbreviations* $\bot, \vee, \rightarrow, \leftrightarrow$ *and* $\Box$ *are defined as usual. For each* $\circ \in \{+, -, \circlearrowleft\}$, *the operator* $[\circ]$ *is defined by* $[\circ] := \neg\langle\circ\rangle\neg$.

In what follows, let Nom be a fixed denumerable set of nominals and we write $\mathcal{L}$ for $\mathcal{L}_{\mathsf{Nom}}$, if there is no danger of confusion.

**Definition 2 (Model).** *A model for* GLV *is a tuple* $\mathfrak{M} = (W, R, V)$, *where* $W$ *is a non-empty set,* $R$ *a binary relation on* $W$ *and* $V :$ Prop $\cup$ Nom $\rightarrow \mathcal{P}(W)$ *a valuation function such that* $V(a)$ *is a singleton set for each* $a \in$ Nom.

Let $\mathfrak{M} = (W, R, V)$ be a model. For each nominal $a$, let $\bar{a}$ denote the point $w \in W$ with $w \in V(a)$. For each $w \in W$, let $R(w) = \{u \in W : Rwu\}$.

**Definition 3 (Truth conditions).** *Let* $\mathfrak{M} = (W, R, V)$ *be a model. Truth of the formula* $\varphi$ *in* $\mathfrak{M}$ *at* $w \in W$ *is defined inductively by:*

$$
\begin{array}{lll}
\mathfrak{M}, w \models x & \textit{iff} & w \in V(x), \textit{ for all } x \in \mathsf{Prop} \cup \mathsf{Nom} \\
\mathfrak{M}, w \models \neg\varphi & \textit{iff} & \mathfrak{M}, w \not\models \varphi \\
\mathfrak{M}, w \models \varphi \wedge \psi & \textit{iff} & \mathfrak{M}, w \models \varphi \textit{ and } \mathfrak{M}, w \models \psi \\
\mathfrak{M}, w \models \Diamond\varphi & \textit{iff} & \mathfrak{M}, u \models \varphi \textit{ for some } u \in W \textit{ such that } Rwu \\
\mathfrak{M}, w \models @_a\varphi & \textit{iff} & \mathfrak{M}, \bar{a} \models \varphi \\
\mathfrak{M}, w \models \langle + \rangle\varphi & \textit{iff} & \textit{there exists } u, v \in W \textit{ such that} \\
& & \langle u, v \rangle \notin R \textit{ and } \mathfrak{M}|_{\langle u+v \rangle}, w \models \psi \\
\mathfrak{M}, w \models \langle - \rangle\varphi & \textit{iff} & \textit{there exists } u, v \in W \textit{ such that} \\
& & \langle u, v \rangle \in R \textit{ and } \mathfrak{M}|_{\langle u-v \rangle}, w \models \psi \\
\mathfrak{M}, w \models \langle \circlearrowleft \rangle\varphi & \textit{iff} & \textit{there exists } u, v \in W \textit{ such that} \\
& & \langle u, v \rangle \in R \textit{ and } \mathfrak{M}|_{\langle u\circlearrowleft v \rangle}, w \models \psi
\end{array}
$$

*where for each* $\circ \in \{+, -, \circlearrowleft\}$, $\mathfrak{M}|_{\langle u\circ v \rangle} = (W, R|_{\langle u\circ v \rangle}, V)$ *is defined by setting*

$$R|_{\langle u+v \rangle} = R \cup \{\langle u, v \rangle\}, \ R|_{\langle u-v \rangle} = R \setminus \{\langle u, v \rangle\}$$

*and*

$$R|_{\langle u\circlearrowleft v \rangle} = \begin{cases} (R|_{\langle u-v \rangle})|_{\langle v+u \rangle}, & \textit{if } \langle u, v \rangle \in R; \\ R, & \textit{otherwise.} \end{cases}$$

*A formula* $\varphi \in \mathcal{L}$ *is valid if* $\mathfrak{M}, w \models \varphi$ *for each model* $\mathfrak{M} = (W, R, V)$ *and* $w \in W$. *Let* GLV *denote the set of all valid formulas in* $\mathcal{L}$.

**Fig. 2.** Updates of link variations for heterogeneous graph in Fig. 1

These dynamic operators characterize the corresponding link variations in directed graphs. An example is given in Fig. 2.

It is shown in [1] that $\langle \circlearrowright \rangle$ cannot be defined by $\langle + \rangle$ and $\langle - \rangle$. Let us go a bit deeper into the operator $\langle \circlearrowright \rangle$. The readers can see that $R|_{\langle u \circlearrowright v \rangle}$ is obtained from $R$ by replacing the link $\langle u, v \rangle$ with $\langle v, u \rangle$. To calculate the set $R|_{\langle u, v \rangle}(w)$ for some given model $\mathfrak{M} = (W, R, V)$ and points $w, u, v \in W$, we have to check if $w \in \{u, v\}$ and $\langle u, v \rangle \in R$. To simplify the discussion and proofs, for all nominals $a, b \in \mathsf{Nom}$, we define the formula $\gamma_{a,b}^{\circlearrowright}$ by $\gamma_{a,b}^{\circlearrowright} := \neg(a \vee b) \vee @_a(\neg \Diamond b \vee b)$. The following proposition explains the intuition behind the formula $\gamma_{a,b}^{\circlearrowright}$.

**Proposition 1.** *Let* $\mathfrak{M} = (W, R, V)$ *be a model,* $a, b \in \mathsf{Nom}$ *and* $w, u \in W$. *Then* $\langle w, u \rangle \in R|_{\langle \bar{a} \circlearrowright \bar{b} \rangle}$ *iff (1)* $Rwu$ *and* $\langle \bar{a}, \bar{b} \rangle \neq \langle w, u \rangle$, *or, (2)* $Ruw$ *and* $\langle \bar{a}, \bar{b} \rangle = \langle u, w \rangle$. *As a corollary,* $\mathfrak{M}, w \models \gamma_{a,b}^{\circlearrowright}$ *if and only if* $R(w) = R|_{\langle a \circlearrowright b \rangle}(w)$.

## 2.2   Global Named Dynamic Operators

With the hybrid operators $@_a$, we can define many useful new operators. Let $a, b \in \mathsf{Nom}$, $\varphi \in \mathcal{L}$. The operators $\langle a + b \rangle$, $\langle a - b \rangle$ and $\langle a \circlearrowright b \rangle$ are defined by:

$$\langle a + b \rangle \varphi := (@_a \Diamond b \wedge \varphi) \vee (@_a \neg \Diamond b \wedge \langle + \rangle(@_a \Diamond b \wedge \varphi));$$
$$\langle a - b \rangle \varphi := (@_a \neg \Diamond b \wedge \varphi) \vee (@_a \Diamond b \wedge \langle - \rangle(@_a \neg \Diamond b \wedge \varphi));$$
$$\langle a \circlearrowright b \rangle \varphi := (@_a(\neg \Diamond b \vee b) \wedge \varphi) \vee (@_a(\Diamond b \wedge \neg b) \wedge \langle \circlearrowright \rangle(@_a \neg \Diamond b \wedge \varphi)).$$

Operators of the form $\langle a + b \rangle$, $\langle a - b \rangle$ and $\langle a \circlearrowright b \rangle$ are called named link adding, cutting and rotating operators, respectively. For each $\circ \in \{+, -, \circlearrowright\}$, $\mathsf{ND}(\circ) = \{\langle a \circ b \rangle : a, b \in \mathsf{Nom}\}$. Let $\mathsf{ND} = \mathsf{ND}(+) \cup \mathsf{ND}(-) \cup \mathsf{ND}(\circlearrowright)$ and $\mathsf{SD} = \bigcup_{n \in \mathbb{N}} \mathsf{ND}^n$. Elements in the set $\mathsf{ND}$ are called named dynamic operators (NDO). The set $\mathsf{SD}$ is consist of all finite sequences of NDOs. For all $\sigma, \delta \in \mathsf{SD}$, we write $\sigma * \delta$ for the concatenation of $\sigma$ and $\delta$. For each $\sigma = \langle s_0, \cdots, s_{n-1} \rangle \in \mathsf{SD}$, we write $\sigma\varphi$ for the formula $s_0 \cdots s_{n-1}\varphi$. For example, if $\sigma = \langle \langle a \circlearrowright b \rangle, \langle c + d \rangle \rangle$ and $\delta = \langle e - f \rangle$, then $\Diamond p \vee \sigma * \delta q = \Diamond p \vee \langle a \circlearrowright b \rangle \langle c + d \rangle \langle e - f \rangle q$.

The following lemma indicates that NDOs do characterize the corresponding model updates, as we desired.

**Lemma 1.** *Let* $\mathfrak{M} = (W, R, V)$ *be a model and* $w \in W$. *Then for all NDOs* $\langle a \circ b \rangle \in \mathsf{ND}$, *and formula* $\varphi \in \mathcal{L}$,

$$\mathfrak{M}, w \models \langle a \circ b \rangle \varphi \text{ if and only if } \mathfrak{M}|_{\langle \bar{a} \circ \bar{b} \rangle}, w \models \varphi.$$

*Proof.* (1) $\circ\ =\ +$. Suppose $\mathfrak{M}, w \models \langle a+b\rangle\varphi$. Then $\mathfrak{M}, w \models (@_a\Diamond b \wedge \varphi) \vee (@_a\neg\Diamond b \wedge \langle+\rangle(@_a\Diamond b \wedge \varphi))$. Suppose $\mathfrak{M}, w \models @_a\Diamond b \wedge \varphi$. Then $\mathfrak{M}, \bar{a} \models \Diamond b$ and $\mathfrak{M}, w \models \varphi$. Then $\langle\bar{a}, \bar{b}\rangle \in R$, which entails $\mathfrak{M} = \mathfrak{M}|_{\langle\bar{a}+\bar{b}\rangle}$ and so $\mathfrak{M}|_{\langle\bar{a}+\bar{b}\rangle}, w \models \varphi$. Suppose $\mathfrak{M}, w \models @_a\neg\Diamond b \wedge \langle+\rangle(@_a\Diamond b \wedge \varphi)$. Then $\mathfrak{M}, \bar{a} \not\models \Diamond b$ and there are $u, v \in W$ with $\mathfrak{M}|_{\langle u+v\rangle}, w \models @_a\Diamond b \wedge \varphi$. Thus $\langle\bar{a}, \bar{b}\rangle \in R|_{\langle u+v\rangle}$ and $\mathfrak{M}|_{\langle u+v\rangle}, w \models \varphi$. Note that $\langle\bar{a}, \bar{b}\rangle \notin R$, we see $\langle\bar{a}, \bar{b}\rangle = \langle u, v\rangle$ and so $\mathfrak{M}|_{\langle\bar{a}+\bar{b}\rangle}, w \models \varphi$.

Suppose $\mathfrak{M}|_{\langle\bar{a}+\bar{b}\rangle}, w \models \varphi$. Assume $\langle\bar{a}, \bar{b}\rangle \in R$. Then $R|_{\langle\bar{a},\bar{b}\rangle} = R$ and so $\mathfrak{M}, w \models @_a\Diamond b \wedge \varphi$, which entails $\mathfrak{M}, w \models \langle a+b\rangle\varphi$. Assume $\langle\bar{a}, \bar{b}\rangle \notin R$. Since $\langle\bar{a}, \bar{b}\rangle \in R|_{\langle\bar{a}+\bar{b}\rangle}$, we have $\mathfrak{M}|_{\langle\bar{a}+\bar{b}\rangle}, w \models @_a\Diamond b$. Since $\mathfrak{M}|_{\langle\bar{a}+\bar{b}\rangle}, w \models \varphi$, we see $\mathfrak{M}, w \models \langle+\rangle(@_a\Diamond b \wedge \varphi)$. Note that $\mathfrak{M}, w \models @_a\neg\Diamond b$, we see $\mathfrak{M}, w \models \langle a+b\rangle\varphi$.

(2) $\circ = -$. Similar to (1).

(3) $\circ = \circlearrowleft$. Suppose $\mathfrak{M}, w \models \langle a \circlearrowleft b\rangle\varphi$. Then $\mathfrak{M}, w \models (@_a(\neg\Diamond b \vee b) \wedge \varphi) \vee (@_a(\Diamond b \wedge \neg b) \wedge \langle\circlearrowleft\rangle(@_a\neg\Diamond b \wedge \varphi))$. Suppose $\mathfrak{M}, w \models @_a(\neg\Diamond b \vee b) \wedge \varphi$. Since $\mathfrak{M}, w \models @_a(\neg\Diamond b \vee b)$, either $\langle\bar{a}, \bar{b}\rangle \notin R$ or $\bar{a} = \bar{b} \in R(\bar{a})$. Thus $R = R|_{\langle\bar{a}\circlearrowleft\bar{b}\rangle}$ and so $\mathfrak{M}|_{\langle\bar{a}\circlearrowleft\bar{b}\rangle}, w \models \varphi$. Suppose $\mathfrak{M}, w \models @_a(\Diamond b \wedge \neg b) \wedge \langle\circlearrowleft\rangle(@_a\neg\Diamond b \wedge \varphi)$. Then we have $\langle\bar{a}, \bar{b}\rangle \in R$, $\bar{a} \neq \bar{b}$ and $\mathfrak{M}, w \models \langle\circlearrowleft\rangle(@_a\neg\Diamond b \wedge \varphi)$. Since $\mathfrak{M}, w \models \langle\circlearrowleft\rangle(@_a\neg\Diamond b \wedge \varphi)$, there exist $u, v \in W$ such that $\langle u, v\rangle \in R$ and $\mathfrak{M}|_{\langle u\circlearrowleft v\rangle}, w \models @_a\neg\Diamond b \wedge \varphi$. Since $\mathfrak{M}|_{\langle u\circlearrowleft v\rangle}, w \models @_a\neg\Diamond b$, we see $\langle\bar{a}, \bar{b}\rangle \notin R|_{\langle u\circlearrowleft v\rangle}$. Note that $\langle\bar{a}, \bar{b}\rangle \in R$, we have $\langle\bar{a}, \bar{b}\rangle = \langle u, v\rangle$ and so $\mathfrak{M}|_{\langle u\circlearrowleft v\rangle} = \mathfrak{M}|_{\langle\bar{a}\circlearrowleft\bar{b}\rangle}$. Hence, $\mathfrak{M}|_{\langle\bar{a}\circlearrowleft\bar{b}\rangle}, w \models \varphi$.

Suppose $\mathfrak{M}|_{\langle\bar{a}\circlearrowleft\bar{b}\rangle}, w \models \varphi$. Assume $\bar{b} \notin R(\bar{a})$ or $\bar{a} = \bar{b}$. Then $R|_{\langle\bar{a}\circlearrowleft\bar{b}\rangle} = R$. Since $\mathfrak{M}, w \models @_a(\neg\Diamond b \vee b)$ and $\mathfrak{M}, w \models \varphi$, $\mathfrak{M}, w \models \langle a \circlearrowleft b\rangle\varphi$. Assume $R\bar{a}\bar{b}$ and $\bar{a} \neq \bar{b}$. Then $\mathfrak{M}, w \models @_a(\Diamond b \wedge \neg b)$. Note that $\langle\bar{a}, \bar{b}\rangle \notin R|_{\langle\bar{a}\circlearrowleft\bar{b}\rangle}$, $\mathfrak{M}|_{\langle\bar{a}\circlearrowleft\bar{b}\rangle}, w \models @_a\neg\Diamond b$. Since $\mathfrak{M}|_{\langle\bar{a}\circlearrowleft\bar{b}\rangle}, w \models \varphi$, we see $\mathfrak{M}, w \models \langle\circlearrowleft\rangle(@_a\neg\Diamond b\wedge\varphi)$. Hence, $\mathfrak{M}, w \models \langle a \circlearrowleft b\rangle\varphi$.

## 3   Axiomatization of Logic of Link Variation

In this section, we introduce a Hilbert-style calculus $\mathsf{C_{GLV}}$ for the logic GLV. Axioms and rules are as follows where $\circ \in \{+, -, \circlearrowleft\}$:

(G1) Axioms and rules for hybrid logic H(@) (cf. [10])
(G2) K-axioms and Necessitation rules for $\langle a+b\rangle$, $\langle a-b\rangle$ and $\langle a \circlearrowleft b\rangle$.
(G3) Axioms and rules for $\langle a+b\rangle$, $\langle a-b\rangle$ and $\langle a \circlearrowleft b\rangle$:

(a) $\langle a \circ b\rangle x \leftrightarrow x$, for $x \in \mathsf{Prop} \cup \mathsf{Nom}$ and $\circ \in \{+, -, \circlearrowleft\}$
(b) $\langle a \circ b\rangle\neg\varphi \leftrightarrow \neg\langle a \circ b\rangle\varphi$, for all $\circ \in \{+, -, \circlearrowleft\}$
(c) $\langle a \circ b\rangle(\varphi \wedge \psi) \leftrightarrow (\langle a \circ b\rangle\varphi \wedge \langle a \circ b\rangle\psi)$, for all $\circ \in \{+, -, \circlearrowleft\}$
(d) $\langle a \circ b\rangle@_i\varphi \leftrightarrow @_i\langle a \circ b\rangle\varphi$, for all $\circ \in \{+, -, \circlearrowleft\}$
(e) $\langle a+b\rangle\Diamond\varphi \leftrightarrow (a \wedge @_b\langle a+b\rangle\varphi) \vee \Diamond\langle a+b\rangle\varphi$
(f) $\langle a-b\rangle\Diamond\varphi \leftrightarrow \Diamond(\neg b \wedge \langle a-b\rangle\varphi) \vee (\neg a \wedge \Diamond\langle a-b\rangle\varphi)$
(g) $\langle a \circlearrowleft b\rangle\Diamond\varphi \leftrightarrow (\gamma_{a,b}^{\circlearrowleft} \wedge \Diamond\langle a \circlearrowleft b\rangle\varphi) \vee (\neg\gamma_{a,b}^{\circlearrowleft} \wedge \psi_{a,b}^{\circlearrowleft})$, where
$\psi_{a,b}^{\circlearrowleft} = (a \wedge \Diamond(\neg b \wedge \langle a \circlearrowleft b\rangle\varphi)) \vee (b \wedge (\Diamond\langle a \circlearrowleft b\rangle\varphi \vee @_a\langle a \circlearrowleft b\rangle\varphi))$
(h) $@_a\neg\Diamond b \wedge \langle a+b\rangle\varphi \rightarrow \langle+\rangle\varphi$
(i) $@_a\Diamond b \wedge \langle a-b\rangle\varphi \rightarrow \langle-\rangle\varphi$
(j) $@_a\Diamond b \wedge \langle a \circlearrowleft b\rangle\varphi \rightarrow \langle\circlearrowleft\rangle\varphi$

(M+) $\dfrac{@_c\sigma(@_a\neg\Diamond b \wedge \langle a+b\rangle\varphi) \to \psi}{@_c\sigma\langle+\rangle\varphi \to \psi}$, where $\sigma \in \mathsf{SD}$ and $a,b$ are new to $\sigma,\varphi,\psi,c$.

(M-) $\dfrac{@_c\sigma(@_a\Diamond b \wedge \langle a-b\rangle\varphi) \to \psi}{@_c\sigma\langle-\rangle\varphi \to \psi}$, where $\sigma \in \mathsf{SD}$ and $a,b$ are new to $\sigma,\varphi,\psi,c$.

(M↻) $\dfrac{@_c\sigma(@_a\Diamond b \wedge \langle a \circlearrowright b\rangle\varphi) \to \psi}{@_c\sigma\langle\circlearrowright\rangle\varphi \to \psi}$, where $\sigma \in \mathsf{SD}$ and $a,b$ are new to $\sigma,\varphi,\psi,c$.

Derivations in $\mathsf{C_{GLV}}$ are defined as usual. For each formula $\varphi$, we write $\vdash \varphi$ if there is a derivation of $\varphi$ in $\mathsf{C_{GLV}}$. In (G3), we provide 'recursion axioms' (a-g) for named dynamic operators. Moreover, axioms (h-j) and the mix-rules show the connections between the named operators and the original dynamic operators.

**Theorem 1 (Soundness).** *For all formula $\varphi \in \mathcal{L}$, $\vdash \varphi$ implies $\varphi \in \mathsf{GLV}$*

*Proof.* We consider only axioms and rules in (G3). Clearly, axioms (G3, a-d) and (G3, h-j) are valid. Validity of (f) is shown in [9], and we will show that axioms (e) and (g) are valid. Let $\mathfrak{M} = (W, R, V)$ be an arbitrary model and $w \in W$.

**(e, $\Rightarrow$)** Suppose $\mathfrak{M}, w \models \langle a+b\rangle\Diamond\varphi$. Then we see $\mathfrak{M}|_{\langle\bar{a}+\bar{b}\rangle}, w \models \Diamond\varphi$ and so there is $v \in R|_{\langle\bar{a}+\bar{b}\rangle}(w)$ with $\mathfrak{M}|_{\langle\bar{a}+\bar{b}\rangle}, v \models \varphi$. By Lemma 1, $\mathfrak{M}, v \models \langle a+b\rangle\varphi$. If $\langle w,v\rangle \in R$, then $\mathfrak{M}, w \models \Diamond\langle a+b\rangle\varphi$. Suppose $\langle w,v\rangle \notin R$. Since $R|_{\langle\bar{a}+\bar{b}\rangle} = R \cup \{\langle\bar{a},\bar{b}\rangle\}$, we see $\langle w,v\rangle = \langle\bar{a},\bar{b}\rangle$. Thus $\mathfrak{M}, w \models a$ and $\mathfrak{M}, v \models b$, which entails $\mathfrak{M}, w \models a \wedge @_b\langle a+b\rangle\varphi$.

**(e, $\Leftarrow$).** Suppose $\mathfrak{M}, w \models a \wedge @_b\langle a+b\rangle\varphi$. Then $\bar{a} = w$ and $\mathfrak{M}, \bar{b} \models \langle a+b\rangle\varphi$. By Lemma 1, $\mathfrak{M}|_{\langle\bar{a}+\bar{b}\rangle}, \bar{b} \models \varphi$. Since $\langle\bar{a},\bar{b}\rangle \in R|_{\langle\bar{a}+\bar{b}\rangle}$, $\mathfrak{M}|_{\langle\bar{a}+\bar{b}\rangle}, w \models \Diamond\varphi$. Thus $\mathfrak{M}, w \models \langle a+b\rangle\Diamond\varphi$. Suppose $\mathfrak{M}, w \models \Diamond\langle a+b\rangle\varphi$. Then there is $v \in R(w)$ with $\mathfrak{M}, v \models \langle a+b\rangle\varphi$. By Lemma 1, $\mathfrak{M}|_{\langle\bar{a}+\bar{b}\rangle}, v \models \varphi$. Since $\langle w,v\rangle \in R \subseteq R|_{\langle\bar{a}+\bar{b}\rangle}$, we have $\mathfrak{M}|_{\langle\bar{a}+\bar{b}\rangle}, w \models \Diamond\varphi$ and so $\mathfrak{M}, w \models \langle a+b\rangle\Diamond\varphi$.

**(g, $\Rightarrow$).** Suppose $\mathfrak{M}, w \models \langle a \circlearrowright b\rangle\Diamond\varphi$. Then $\mathfrak{M}|_{\langle\bar{a}\circlearrowright\bar{b}\rangle}, w \models \Diamond\varphi$, which entails $\mathfrak{M}|_{\langle\bar{a}\circlearrowright\bar{b}\rangle}, v \models \varphi$ for some $v \in R|_{\langle\bar{a}\circlearrowright\bar{b}\rangle}(w)$. By Lemma 1, $\mathfrak{M}, v \models \langle a \circlearrowright b\rangle\varphi$. Suppose $\mathfrak{M}, w \models \gamma_{a,b}^{\circlearrowright}$. By Proposition 1, $R(w) = R|_{\langle a\circlearrowright b\rangle}(w)$ and so $v \in R(w)$. Thus $\mathfrak{M}, w \models \gamma_{a,b}^{\circlearrowright} \wedge \Diamond\langle a \circlearrowright b\rangle\varphi$. Suppose $\mathfrak{M}, w \not\models \gamma_{a,b}^{\circlearrowright}$. Then $\mathfrak{M}, w \models @_a(\Diamond b \wedge \neg b) \wedge (a \vee b)$, which entails $R\bar{a}\bar{b}$, $\bar{a} \neq \bar{b}$ and $w \in \{\bar{a}, \bar{b}\}$. Since $R\bar{a}\bar{b}$, we see $R|_{\langle\bar{a}\circlearrowright\bar{b}\rangle} = (R \setminus \{\langle\bar{a},\bar{b}\rangle\}) \cup \{\langle\bar{b},\bar{a}\rangle\}$. Now we have two cases:

(1) $w = \bar{a}$. Since $\bar{a} \neq \bar{b}$, we see $R|_{\langle a\circlearrowright b\rangle}(w) = R(w)\setminus\{\bar{b}\}$. Since $v \in R|_{\langle a\circlearrowright b\rangle}(w)$, we see $v \neq \bar{b}$ and $Rwv$, which entails $\mathfrak{M}, w \models \Diamond(\neg b \wedge \langle a \circlearrowright b\rangle\varphi)$.

(2) $w = \bar{b}$. Since $\bar{a} \neq \bar{b}$ and $v \in R|_{\langle a\circlearrowright b\rangle}(w)$, we see $v = \bar{a}$ or $Rwv$. If $v = \bar{a}$, then $\mathfrak{M}, w \models @_a\langle a \circlearrowright b\rangle\varphi$. If $Rwv$, then $\mathfrak{M}, w \models \Diamond\langle a \circlearrowright b\rangle\varphi$.

In both of these cases, we see $\mathfrak{M}, w \models \psi_{a,b}^{\circlearrowright}$.

**(g, $\Leftarrow$).** The proof proceeds by the following two parts:

(1) $\models \gamma_{a,b}^{\circlearrowright} \wedge \Diamond\langle a \circlearrowright b\rangle\varphi \to \langle a \circlearrowright b\rangle\Diamond\varphi$. Suppose $\mathfrak{M}, w \models \gamma_{a,b}^{\circlearrowright} \wedge \Diamond\langle a \circlearrowright b\rangle\varphi$. Since $\mathfrak{M}, w \models \Diamond\langle a \circlearrowright b\rangle\varphi$, there exists $v \in R(w)$ such that $\mathfrak{M}, v \models \langle a \circlearrowright b\rangle\varphi$. By Lemma 1, $\mathfrak{M}|_{\langle\bar{a}\circlearrowright\bar{b}\rangle}, v \models \varphi$. Since $\mathfrak{M}, w \models \gamma_{a,b}^{\circlearrowright}$, by Proposition 1, $R(w) = R|_{\langle a\circlearrowright b\rangle}(w)$. Thus $v \in R|_{\langle a\circlearrowright b\rangle}(w)$, which entails $\mathfrak{M}|_{\langle a\circlearrowright b\rangle}, w \models \Diamond\varphi$. By Lemma 1, $\mathfrak{M}, w \models \langle a \circlearrowright b\rangle\Diamond\varphi$.

(2) $\models \neg\gamma_{a,b}^{\circlearrowright} \wedge \psi_{a,b}^{\circlearrowright} \rightarrow \langle a \circlearrowright b \rangle \Diamond\varphi$, where

$$\psi_{a,b}^{\circlearrowright} = (a \wedge \Diamond(\neg b \wedge \langle a \circlearrowright b \rangle \varphi)) \vee (b \wedge (\Diamond\langle a \circlearrowright b \rangle \varphi \vee @_a \langle a \circlearrowright b \rangle \varphi)).$$

Suppose $\mathfrak{M}, w \models \neg\gamma_{a,b}^{\circlearrowright} \wedge \psi_{a,b}^{\circlearrowright}$. Since $\mathfrak{M}, w \models \neg\gamma_{a,b}^{\circlearrowright}$, we have $R\bar{a}\bar{b}$, $\bar{a} \neq \bar{b}$ and $w \in \{\bar{a}, \bar{b}\}$. Since $R\bar{a}\bar{b}$, $R|_{\langle \bar{a} \circlearrowright \bar{b} \rangle} = (R \setminus \{\langle \bar{a}, \bar{b} \rangle\}) \cup \{\langle \bar{b}, \bar{a} \rangle\}$. Now we have two cases:

(2.1) $w = \bar{a}$. Then $\mathfrak{M}, w \not\models b$ and $R|_{\langle a \circlearrowright b \rangle}(w) = R(w) \setminus \{b\}$. Since $\mathfrak{M}, w \models \psi_{a,b}^{\circlearrowright}$, we see $\mathfrak{M}, w \models a \wedge \Diamond(\neg b \wedge \langle a \circlearrowright b \rangle \varphi)$. Then there exists $v \in R(w)$ with $\mathfrak{M}, v \models \neg b \wedge \langle a \circlearrowright b \rangle \varphi$. By Lemma 1, $\mathfrak{M}|_{\langle \bar{a} \circlearrowright \bar{b} \rangle}, v \models \varphi$. Since $\bar{b} \neq v \in R(w)$, $v \in R|_{\langle a \circlearrowright b \rangle}(w)$ and so $\mathfrak{M}|_{\langle a \circlearrowright b \rangle}, w \models \Diamond\varphi$. By Lemma 1, $\mathfrak{M}, w \models \langle a \circlearrowright b \rangle \Diamond\varphi$.

(2.2) $w = \bar{b}$. Then $\mathfrak{M}, w \not\models a$ and $R|_{\langle a \circlearrowright b \rangle}(w) = R(w) \cup \{\bar{a}\}$. Since $\mathfrak{M}, w \models \psi_{a,b}^{\circlearrowright}$, we see $\mathfrak{M}, w \models b \wedge (\Diamond\langle a \circlearrowright b \rangle \varphi \vee @_a \langle a \circlearrowright b \rangle \varphi)$. Suppose $\mathfrak{M}, w \models \Diamond\langle a \circlearrowright b \rangle \varphi$, then there exists $v \in R(w)$ such that $\mathfrak{M}, v \models \neg b \wedge \langle a \circlearrowright b \rangle \varphi$. By Lemma 1, $\mathfrak{M}|_{\langle \bar{a} \circlearrowright \bar{b} \rangle}, v \models \varphi$. Note that $v \in R(w) \subseteq R|_{\langle a \circlearrowright b \rangle}(w)$, we see $\mathfrak{M}|_{\langle a \circlearrowright b \rangle}, w \models \Diamond\varphi$ and so $\mathfrak{M}, w \models \langle a \circlearrowright b \rangle \Diamond\varphi$. Suppose $\mathfrak{M}, w \models @_a \langle a \circlearrowright b \rangle \varphi$. Then $\mathfrak{M}|_{\langle a \circlearrowright b \rangle}, \bar{a} \models \varphi$. Since $\langle w, \bar{a} \rangle \in R|_{\langle a \circlearrowright b \rangle}$, we see $\mathfrak{M}|_{\langle a \circlearrowright b \rangle}, w \models \Diamond\varphi$ and so $\mathfrak{M}, w \models \langle a \circlearrowright b \rangle \Diamond\varphi$.

## 4    Completeness for $\mathsf{C_{GLV}}$

The aim of this section is to show the completeness for $\mathsf{C_{GLV}}$ with respect to GLV. The sketch of our proof is as follows: Let $\Gamma \subseteq \mathcal{L}$ be an arbitrarily fixed named, pasted and mixed maximal consistent set of formulas. Then we construct a family of canonical models based on $\Gamma$ and show that these canonical models characterize the behaviour of the dynamic operators. Finally, we show that $\Gamma$ is satisfied by one of those models. Note that every consistent set of formulas can be extended to a named, pasted and mixed MCS in some properly extended formal language, we are done.

**Definition 4.** *Let $\mathcal{L}$ be a language and $\Gamma \subseteq \mathcal{L}$ a set of formulas. Then we say*

- *$\Gamma$ is consistent, if $\nvdash \varphi_1 \wedge \cdots \wedge \varphi_n \rightarrow \bot$ for any $\varphi_1, \cdots, \varphi_n \in \Gamma$.*
- *$\Gamma$ is $\mathcal{L}$-maximal consistent, if $\Gamma$ is consistent and $\Delta \vdash \bot$ for all $\Gamma \subsetneq \Delta \subseteq \mathcal{L}$.*
- *$\Gamma$ is named, if $i \in \Gamma$ for some nominal $i \in \mathsf{Nom}$.*
- *$\Gamma$ is pasted, if for all $@_i \Diamond\varphi \in \Gamma$, there is $j \in \mathsf{Nom}$ such that $@_i \Diamond j \wedge @_j \varphi \in \Gamma$.*
- *$\Gamma$ is mixed, if for all $i \in \mathsf{Nom}$, $\sigma \in \mathsf{SD}$ and $\varphi \in \mathcal{L}$, we have:*
  - *if $@_i \sigma \langle + \rangle \varphi \in \Gamma$, then $@_i \sigma (@_a \neg \Diamond b \wedge \langle a + b \rangle \varphi) \in \Gamma$ for some $a, b \in \mathsf{Nom}$.*
  - *if $@_i \sigma \langle - \rangle \varphi \in \Gamma$, then $@_i \sigma (@_a \Diamond b \wedge \langle a - b \rangle \varphi) \in \Gamma$ for some $a, b \in \mathsf{Nom}$.*
  - *if $@_i \sigma \langle \circlearrowright \rangle \varphi \in \Gamma$, then $@_i \sigma (@_a \Diamond b \wedge \langle a \circlearrowright b \rangle \varphi) \in \Gamma$ for some $a, b \in \mathsf{Nom}$.*

*$\Gamma$ is called an $\mathcal{L}$-maximal consistent set ($\mathcal{L}$-MCS) if it is $\mathcal{L}$-maximal consistent.*

**Lemma 2.** *Let $\mathcal{L}'$ be a language obtained by extending $\mathcal{L}$ with a denumerable set $\mathsf{Nom}_0$ of new nominals. Then every $\mathcal{L}$-consistent set can be extended to a named, pasted and mixed $\mathcal{L}'$-MCS.*

*Proof.* Let $\Gamma$ be an $\mathcal{L}$-consistent set and $\Gamma_0 = \Gamma \cup \{j_0\}$. By the rule (Name), one can readily check that $\Gamma_0$ is $\mathcal{L}'$-consistent. Let $(\varphi_n)_{n \in \mathbb{N}}$ be an enumeration of all formulas in $\mathcal{L}'$. Then for each $k \in \mathbb{N}$, we define the set $\Gamma_{k+1}$ as follows: If $\Gamma_k \cup \{\varphi_k\}$ is $\mathcal{L}'$-inconsistent, then $\Gamma_{k+1} = \Gamma_k$. Otherwise,

- $\Gamma_{k+1} = \Gamma_k \cup \{\varphi_k\} \cup \{@_i \Diamond j \wedge @_j \psi\}$ if $\varphi_k$ is of the form $@_i \Diamond \psi$,
  where $j \in \mathsf{Nom}_0$ is the first new nominal w.r.t $\Gamma_k$ and $\varphi_k$.
- $\Gamma_{k+1} = \Gamma_k \cup \{\varphi_k\} \cup \{@_i \sigma (@_a \neg \Diamond b \wedge \langle a+b \rangle \psi)\}$ if $\varphi_k$ is of the form $@_i \sigma \langle + \rangle \psi$,
  where $a, b \in \mathsf{Nom}_0$ are new nominals w.r.t $\Gamma_k$ and $\varphi_k$.
- $\Gamma_{k+1} = \Gamma_k \cup \{\varphi_k\} \cup \{@_i \sigma (@_a \Diamond b \wedge \langle a \circ b \rangle \psi)\}$ if $\varphi_k$ is of the form $@_i \sigma \langle \circ \rangle \psi$
  ($\circ \in \{-, \circlearrowright\}$),where $a, b \in \mathsf{Nom}_0$ are new nominals w.r.t $\Gamma_k$ and $\varphi_k$.
- $\Gamma_{k+1} = \Gamma_k \cup \{\varphi_k\}$ if $\varphi_k$ is not of the above forms.

In the construction above, by new nominals we mean those with minimal index in $\mathsf{Nom}_0$. Let $\Gamma^* = \bigcup_{n \in \mathbb{N}} \Gamma_n$. By the rules (Paste), (M+), (M−) and (M$\circlearrowright$), $\Gamma_k$ is consistent for all $k \in \mathbb{N}$. Then one can readily check that $\Gamma^*$ is what we desired.

We now show the strong completeness of the calculus $\mathsf{C}_{\mathsf{GLV}}$. Let $\Gamma$ be a fixed consistent set of $\mathcal{L}$-formulas. Since $\mathcal{L}$ contains already a denumerable set of nominals, by Lemma 2, we can assume that $\Gamma$ itself is a named, pasted and mixed $\mathcal{L}$-MCS. It suffices to show that $\Gamma$ is satisfiable.

**Definition 5.** *For each $j \in \mathsf{Nom}$, let $\Delta_j = \{\varphi : @_j \varphi \in \Gamma\}$. Then the canonical model induced by $\Gamma$ is the tuple $\mathfrak{M}^\Gamma = \langle W^\Gamma, R^\Gamma, V^\Gamma \rangle$, where*

- $W^\Gamma = \{\Delta_i : i \in \mathsf{Nom}\}$.
- $R^\Gamma \Delta_i \Delta_j$ *iff* $@_i \Diamond j \in \Gamma$.
- $V^\Gamma(x) = \{w \in W^\Gamma : x \in w\}$, *for all* $x \in \mathsf{Prop} \cup \mathsf{Nom}$.

*Moreover, for each sequence $\sigma \in \mathsf{SD}$, the $\sigma$-canonical model $\mathfrak{M}^\sigma = (W^\sigma, R^\sigma, V^\sigma)$ induced by $\Gamma$ is defined inductively as follows:*

- $W^\sigma = \{w^\sigma : w \in W^\Gamma\}$, *where* $w^\sigma = \{\varphi : \sigma \varphi \in w\}$.
- $R^\epsilon = R^\Gamma$.
- *If* $\sigma = \sigma' * \langle a+b \rangle$, *then* $R^\sigma w^\sigma v^\sigma$ *iff* $R^{\sigma'} w^{\sigma'} v^{\sigma'}$ *or* $\langle a, b \rangle \in w \times v$.
- *If* $\sigma = \sigma' * \langle a-b \rangle$, *then* $R^\sigma w^\sigma v^\sigma$ *iff* $R^{\sigma'} w^{\sigma'} v^{\sigma'}$ *and* $\langle a, b \rangle \notin w \times v$.
- *If* $\sigma = \sigma' * \langle a \circlearrowright b \rangle$, *then* $R^\sigma w^\sigma v^\sigma$ *iff one of the following conditions holds:*
  *(1)* $R^{\sigma'} w^{\sigma'} v^{\sigma'}$ *and* $\langle a, b \rangle \notin w \times v$.
  *(2)* $R^{\sigma'} v^{\sigma'} w^{\sigma'}$ *and* $\langle b, a \rangle \in w \times v$.
- $V^\sigma(x) = \{w^\sigma \in W^\sigma : x \in w\}$, *for all* $x \in \mathsf{Prop} \cup \mathsf{Nom}$.

A family of canonical models are provided in Definition 5. These models are proposed to 'simulate' the dynamic updates syntactically. Before going into the details of the proof of Completeness theorem, let us introduce some basic properties of the canonical models.

**Proposition 2.** *Let $\varphi \in \mathcal{L}$, $a, b \in \mathsf{Nom}$, $\sigma, \delta \in \mathsf{SD}$ and $w = \Delta_b \in W^\Gamma$. Then*

*(1) $w^\sigma$ is an $\mathcal{L}$-MCS.*
*(2) $\delta \varphi \in w^\sigma$ if and only if $\varphi \in w^{\sigma * \delta}$.*
*(3) $@_a \varphi \in w^\sigma$ if and only if $\varphi \in (\Delta_a)^\sigma$.*

*Proof.* For (1), we show first that $w^\sigma$ is consistent. Suppose there are formulas $\psi_1, \cdots, \psi_n \in w^\sigma$ such that $\vdash \psi_1 \wedge \cdots \wedge \psi_n \to \bot$. Then $@_b \sigma \psi_1 \wedge \cdots \wedge @_b \sigma \psi_n \in \Gamma$. By (G2), we have $\vdash @_b \sigma \psi_1 \wedge \cdots \wedge @_b \sigma \psi_n \to @_b \sigma \bot$, which entails $@_b \sigma \bot \in \Gamma$. By axiom (G3, a-c), $@_b \bot \in \Gamma$, which entails $\bot \in \Gamma$ and contradicts the assumption. To show that $w^\sigma$ is $\mathcal{L}$-maximal, it suffices to show $\varphi \notin w^\sigma$ implies $\neg\varphi \in w^\sigma$. Suppose $\varphi \notin w^\sigma$. Then $@_b \sigma \varphi \notin \Gamma$ and so $\neg @_b \sigma \varphi \in \Gamma$. By axiom (G3, b) and (G1), $@_b \sigma \neg\varphi \in \Gamma$, which entails $\neg\varphi \in w^\sigma$.

For (2), note that $\delta\varphi \in w^\sigma$ iff $\sigma\delta\varphi \in w$ iff $\varphi \in w^{\sigma*\delta}$, we are done.

For (3), suppose $w = \Delta_b$. Then by (G3, d) and (G1) we see

$$@_a \varphi \in w^\sigma \text{ iff } @_b \sigma @_a \varphi \in \Gamma \text{ iff } @_a \sigma \varphi \in \Gamma \text{ iff } \varphi \in (\Delta_a)\sigma.$$

Proposition 2 is frequently applied in proofs of the following lemmas and theorems. For example, we will conclude $\psi \in w^\sigma$ from $\vdash \varphi \to \psi$ and $\varphi \in w^\sigma$. We do not specify the application of Proposition 2 in what follows.

**Lemma 3.** *For all NDOs $\langle a \circ b \rangle \in \mathsf{ND}$ and $\sigma \in \mathsf{SD}$, we define the function $f : W^\sigma \to W^{\sigma*\langle a \circ b \rangle}$ by $f : w^\sigma \mapsto w^{\sigma*\langle a \circ b \rangle}$. Then*

$$f : \mathfrak{M}^\sigma|_{\langle \bar{a} \circ \bar{b} \rangle} \cong \mathfrak{M}^{\sigma*\langle a \circ b \rangle}$$

.

**Lemma 4.** *Let $\sigma \in \mathsf{SD}$, $w, v \in W^\Gamma$. Then for all $\varphi \in \mathcal{L}$,*

*(1) if $R^\sigma w^\sigma v^\sigma$ and $\varphi \in v^\sigma$, then $\Diamond\varphi \in w^\sigma$.*
*(2) if $\Diamond\varphi \in w^\sigma$, then there exists $v \in W^\Gamma$ with $R^\sigma w^\sigma v^\sigma$ and $\varphi \in v^\sigma$.*

The proofs of Lemma 3 and Lemma 4 are given in the Appendix. These two lemmas show that the operator $\Diamond$ behave as we desired in all canonical models.

**Lemma 5 (Truth Lemma).** *Let $\psi \in \mathcal{L}$, $w \in W^\Gamma$ and $\sigma \in \mathsf{SD}$. Then*

$$\mathfrak{M}^\sigma, w^\sigma \models \varphi \text{ if and only if } \varphi \in w^\sigma.$$

*Proof.* The proof proceeds by induction on the complexity of $\varphi$.

(1) $\varphi \in \mathsf{Prop} \cup \mathsf{Nom}$. By axiom (G3, a), $\vdash \varphi \leftrightarrow \sigma\varphi$. Then we have

$$\mathfrak{M}^\sigma, w^\sigma \models \varphi \text{ iff } w \in V^\Gamma(\varphi) \text{ iff } \varphi \in w \text{ iff } \sigma\varphi \in w \text{ iff } \varphi \in w^\sigma$$

(2) Boolean and @ cases are taken cared by axioms (G3, b-d).
(3) $\varphi$ is of the form $\Diamond\psi$. Then we see

$$\begin{aligned} \Diamond\psi \in w^\sigma &\text{ iff } R^\sigma w^\sigma v^\sigma \text{ and } \psi \in v^\sigma \text{ for some } v \in W^\Gamma &\text{(Lemma 4)} \\ &\text{ iff } R^\sigma w^\sigma v^\sigma \text{ and } \mathfrak{M}^\sigma, v^\sigma \models \psi \text{ for some } v \in W^\Gamma &\text{(IH)} \\ &\text{ iff } \mathfrak{M}^\sigma, w^\sigma \models \Diamond\psi. \end{aligned}$$

(4) $\varphi$ is of the form $\langle+\rangle\gamma$. Suppose $\mathfrak{M}^\sigma, w^\sigma \models \langle+\rangle\gamma$. Then there are $u^\sigma, v^\sigma \in W^\sigma$ with $\langle u^\sigma, v^\sigma\rangle \notin R^\sigma$ and $\mathfrak{M}^\sigma|_{\langle u^\sigma+v^\sigma\rangle}, w^\sigma \models \gamma$. Let $a, b \in \mathsf{Nom}$ be nominals such that $\langle \bar{a}, \bar{b}\rangle = \langle u^\sigma, v^\sigma\rangle$. Then by Lemma 3, $\mathfrak{M}^{\sigma'}, w^{\sigma'} \models \gamma$. By IH, $\gamma \in w^{\sigma'}$ and so $\langle a+b\rangle\gamma \in w^\sigma$. Since $\langle u^\sigma, v^\sigma\rangle \notin R^\sigma$, by Lemma 4(2), $\Diamond b \notin u^\sigma$, which entails $@_a\Diamond b \notin w^\sigma$. By (G3,h), $\langle+\rangle\gamma \in w^\sigma$. Suppose $\langle+\rangle\gamma \in w^\sigma$. Since $\Gamma$ is mixed, we see there are $a, b \in \mathsf{Nom}$ such that $@_a\neg\Diamond b \wedge \langle a+b\rangle\gamma \in w^\sigma$. Then $@_a\neg\Diamond b \in w^\sigma$ and $\gamma \in w^{\sigma'}$. Since $\gamma \in w^{\sigma'}$, by IH, $\mathfrak{M}^{\sigma'}, w^{\sigma'} \models \gamma$. By Lemma 3, $\mathfrak{M}^\sigma|_{\langle a+b\rangle}, w^\sigma \models \gamma$. Since $@_a\neg\Diamond b \in w^\sigma$, $\Diamond b \notin u^\sigma$. By Lemma 4(1), $\langle u^\sigma, v^\sigma\rangle \notin R^\sigma$. Hence $\mathfrak{M}^\sigma, w^\sigma \models \langle+\rangle\gamma$.

(5) $\varphi$ is of the form $\langle-\rangle\gamma$ or $\langle\circlearrowleft\rangle\gamma$. The proof for this case is similar to (4), where axioms (G3, i-j) are applied. Details are omitted to save space.

By Lemma 5 and the arbitrariness of $\Gamma$, we see

**Theorem 2.** $\mathsf{C_{GLV}}$ *is strongly complete w.r.t.* GLV.

Based on the given definitions and proofs, it is evident that by restricting formal language $\mathcal{L}$ to any subset $X$ of $\{+, -, \circlearrowleft\}$, we can still achieve a sound and complete axiomatization by just ignoring the operators that do not occur in $X$. Let $X \subseteq \{+, -, \circlearrowleft\}$. Then we define

$$\mathcal{L}(X) = \{\varphi \in \mathcal{L} : \text{dynamic operators occur in } \varphi \text{ is among } X\}.$$

Let $\mathsf{GLV}(X) = \mathsf{GLV} \cap \mathcal{L}(X)$ and $\mathsf{C_{GLV}}(X)$ be the calculus consist of axioms and rules from $\mathsf{C_{GLV}}$ in $\mathcal{L}(X)$. $\mathsf{GLV}(X)$ and $\mathsf{C_{GLV}}(X)$ are called $X$-fragment of GLV and $\mathsf{C_{GLV}}$, respectively. Then the following theorem holds:

**Theorem 3.** $\mathsf{C_{GLV}}(X)$ *is sound and strongly complete w.r.t.* $\mathsf{GLV}(X)$.

## 5 Logic of Local Definable Link Variation

In the sections above, we consider the global link variations which involves one link in each action, investigate their logics and provide complete axiomatizations for them. As a variant of global link cutting operation, [12] suggests another kind of link cutting operation, local (definable) link cutting, where a set of links connected to the current point are involved simultaneously. With the results on global link variations, it is natural to investigate the local definable operations for link variations and their logics.

In [12], the definable link cutting operator $[-\varphi]$ is introduced and its logic LLD is investigated. Models of LLD are usual Kripke models, and the truth condition of $[-\psi]\varphi$ is given by:

$$(W, R, V), w \models [-\psi]\varphi \text{ if and only if } (W, R|_{\langle w-\psi\rangle}, V), w \models \varphi,$$

where $R|_{\langle w-\psi\rangle} = R \setminus (\{w\} \times [\![\psi]\!])$.[1] Intuitively, the operation $[-\psi]$ cuts the links between the current point and the points where $\psi$ holds. There are results

---

[1] $[\![\psi]\!] = \{u \in W : \mathfrak{M}, u \models \psi\}$.

on expressive power of LLD, but the model checking problem is still open. Moreover, there is no sound and complete axiomatization for LLD, even for $\mathsf{LLD}(@, \downarrow)$, the logic obtained by adding hybrid operators $\downarrow$ and $@$ to LLD.

In what follows, we extend the logic LLD to the logic LLV of local link variations by adding local adding operator $[+\psi]$, local rotating operator $[\circlearrowright\psi]$ and hybrid operators $\downarrow$ and $\mathsf{E}$. By providing recursion axioms for the local dynamic operators, we obtain a sound and strongly complete axiomatization for LLV.

The language $\mathcal{L}^l$ of LLV is given by

$$\mathcal{L}^l \ni \varphi ::= a \mid p \mid \neg\varphi \mid \varphi \wedge \varphi \mid \Diamond\varphi \mid \mathsf{E}\varphi \mid \downarrow a.\varphi \mid [+\varphi]\varphi \mid [-\varphi]\varphi \mid [\circlearrowright\varphi]\varphi$$

where $a \in \mathsf{Nom}$ and $p \in \mathsf{Prop}$. For each $a \in \mathsf{Nom}$ and $\varphi \in \mathcal{L}^l$, we define $@_a\varphi := \mathsf{E}(a \wedge \varphi)$. Models of LLV are exactly models for the hybrid logic $\mathsf{LLD}(\downarrow, \mathsf{E})$ and semantics of the operators $\downarrow a.$ and $\mathsf{E}$ (global existential operator) are as usual (cf. [10]). Let $\mathfrak{M} = (W, R, V)$ be a model of LLV. For each $w \in W$ and $\psi \in \mathcal{L}^l$, we set $R(w, \psi) = \{w\} \times (\llbracket\psi\rrbracket \cap R(w))$ and $R^{-1}(w, \psi) = \{\langle u, w\rangle : \langle w, u\rangle \in R(w, \psi)\}$. For $\circ \in \{+, \circlearrowright\}$, truth for the formula $[\circ\psi]\varphi$ is given by:

$$\mathfrak{M}, w \models [\circ\psi]\varphi \text{ if and only if } \mathfrak{M}|_{\langle w\circ\psi\rangle}, w \models \varphi$$

where $R|_{\langle w+\psi\rangle} = R \cup (\{w\} \times \llbracket\psi\rrbracket)$ and $R|_{\langle w\circlearrowright\psi\rangle} = (R \setminus R(w, \psi)) \cup R^{-1}(w, \psi)$.

As in Sect. 2, for each $\circ \in \{+, -, \circlearrowright\}$, local named dynamic operators are defined as follows: Let $a \in \mathsf{Nom}$, $\varphi \in \mathcal{L}^l$, the operator $\langle a \circ \varphi\rangle$ is defined by:

$$\langle a \circ \varphi\rangle\psi := \downarrow b.@_a[\circ\varphi]@_b\psi,$$

where $b$ is a new nominal with respect to $\varphi$, $\psi$ and $a$. One should note that the downarrow operator $\downarrow$ is crucial in this definition: the operators $\langle a \circ \varphi\rangle$ allow us change links 'globally', and we have to 'go back to' the original point. If a formula $\varphi \in \mathcal{L}^l$ is of the form where only named operators occur, we call it a named formula. For example, $\downarrow a.@_b[+p]@_i\Diamond q \wedge r$ is a named formula and $[+p]@_a\Diamond q \wedge r$ is not. Let $\mathcal{L}^n$ denote the set of all named formulas.

**Lemma 6.** *Let $\varphi \in \mathcal{L}^l$ and $a, b \in \mathsf{Nom}$ nominals such that $b$ does not occur in $\{\varphi, a\}$. Then for all model $\mathfrak{M} = (W, R, V)$, $w, u \in W$ and $\circ \in \{+, -, \circlearrowright\}$,*

*(1) $\mathfrak{M}, w \models @_a\varphi$ if and only if $\mathfrak{M}, \bar{a} \models \varphi$.*

*(2) $\mathfrak{M}|_b^w, u \models \varphi$ if and only if $\mathfrak{M}, u \models \varphi$.*

*(3) $(\mathfrak{M}|_b^w)|_{\langle a\circ\varphi\rangle} \cong (\mathfrak{M}|_{\langle a\circ\varphi\rangle})|_b^w$.*

*Proof.* The proof of (1) is trivial. For (2), the proof proceeds by induction on the complexity of $\varphi$. (3) follows from (2) immediately.

As the following lemma shows, the local named dynamic operators behave as we desired:

**Lemma 7.** *Let $\mathfrak{M} = (W, R, V)$ be a model and $w \in W$. Then for each $a \in \mathsf{Nom}$, $\circ \in \{+, -, \circlearrowright\}$ and $\varphi, \psi \in \mathcal{L}^l$,*

$$\mathfrak{M}, w \models \langle a \circ \psi \rangle \varphi \text{ if and only if } \mathfrak{M}|_{\langle \bar{a} \circ \psi \rangle}, w \models \varphi.$$

*Proof.* Note that $\langle a \circ \varphi \rangle \psi := \downarrow b.@_a[\circ \varphi]@_b \psi$ for some nominal $b \in \mathsf{Nom}$ which does not occur in $\{\varphi, \psi, a\}$, we have

$$
\begin{aligned}
\mathfrak{M}, w \models \langle a \circ \psi \rangle \varphi &\text{ iff } \mathfrak{M}, w \models \downarrow b.@_a[\circ \psi]@_b \varphi \\
&\text{ iff } (\mathfrak{M}|_b^w)|_{\langle \bar{a} \circ \psi \rangle}, w \models \varphi \quad \text{Lemma 6(1)} \\
&\text{ iff } (\mathfrak{M}|_{\langle \bar{a} \circ \psi \rangle})|_b^w, w \models \varphi \quad \text{Lemma 6(3)} \\
&\text{ iff } \mathfrak{M}|_{\langle \bar{a} \circ \psi \rangle}, w \models \varphi \quad \text{Lemma 6(2)}
\end{aligned}
$$

In what follows, we introduce the calculus $\mathsf{C_{LLV}}$ for the logic LLV. Axioms and rules are as follows where $\circ \in \{+, -, \circlearrowleft\}$:

(L1) Axioms and rules for hybrid logic $\mathsf{H(E, \downarrow)}$ (cf. [10])
(L2) Axioms for local named dynamic operators: for all $\circ \in \{+, -, \circlearrowleft\}$,
  (a) $\langle a \circ \psi \rangle x \leftrightarrow x$, for $x \in \mathsf{Prop} \cup \mathsf{Nom}$
  (b) $\langle a \circ \psi \rangle \neg \varphi \leftrightarrow \neg \langle a \circ \psi \rangle \varphi$
  (c) $\langle a \circ \psi \rangle (\varphi \wedge \psi) \leftrightarrow \langle a \circ \psi \rangle \varphi \wedge \langle a \circ \psi \rangle \psi$
  (d) $\langle a \circ \psi \rangle \downarrow b.\varphi \leftrightarrow \downarrow c.\langle a \circ \psi \rangle (\varphi[b := c])$
  (e) $\langle a + \psi \rangle \Diamond \varphi \leftrightarrow (\Diamond \langle a + \psi \rangle \varphi \vee (a \wedge \mathsf{E}(\psi \wedge \langle a + \psi \rangle \varphi)))$
  (f) $\langle a - \psi \rangle \Diamond \varphi \leftrightarrow ((a \wedge \Diamond(\neg \psi \wedge \langle a - \psi \rangle \varphi)) \vee (\neg a \wedge \Diamond \langle a - \psi \rangle \varphi))$
  (g) $\begin{aligned}[t]\langle a \circlearrowleft \psi \rangle \Diamond \varphi \leftrightarrow \quad &(a \wedge (\Diamond(\neg \psi \wedge \langle a \circlearrowleft \psi \rangle \varphi) \vee (\Diamond a \wedge \langle a \circlearrowleft \psi \rangle \varphi))) \\ &\vee (\neg a \wedge (\Diamond \langle a \circlearrowleft \psi \rangle \varphi \vee (\psi \wedge \downarrow c.@_a \Diamond c \wedge @_a \langle a \circlearrowleft \psi \rangle \varphi)))\end{aligned}$
  (h) $[\circ \psi] \varphi \leftrightarrow \downarrow c.\langle c \circ \psi \rangle \varphi$

In (L2), $\circ$ ranges among $\{+, -, \circlearrowleft\}$ and $c$ is always new to the other formulas. Validity of (L2, d) and (L2, h) is can be easily verified. Similar to Theorem 1, the readers can verify that all axioms in (L2) are valid. As in Sect. 4, for each $X \subseteq \{+, -, \circlearrowleft\}$, let $\mathcal{L}^l(X)$, $\mathcal{L}^n(X)$, $\mathsf{LLV}(X)$ and $\mathsf{C_{LLV}}(X)$ be the $X$-fragment of $\mathcal{L}^l$, $\mathcal{L}^n$, LLV and $\mathsf{C_{LLV}}$, respectively. Specially, we have $\mathsf{LLV}(\varnothing) = \mathsf{H(\downarrow, E)}$.

**Proposition 3.** *Let $X \subseteq \{+, -, \circlearrowleft\}$. Then*

*(1) For all $\varphi \in \mathcal{L}^l(X)$, there is $\varphi' \in \mathcal{L}^n(X)$ such that $\models \varphi \leftrightarrow \varphi^*$.*
*(2) For all $\varphi \in \mathcal{L}^n(X)$, there is $\varphi' \in \mathcal{L}^l(\varnothing)$ such that $\models \varphi \leftrightarrow \varphi'$.*

*Proof.* For (1), we define a translation $(\cdot)^*$ from $\mathcal{L}^l(X)$ to $\mathcal{L}^n$ as follows:

$$
\begin{aligned}
x^* &= x \text{ for all } x \in \mathsf{Prop} \cup \mathsf{Nom} \\
(\circ \varphi)^* &= \circ \varphi^* \text{ for all } \circ \in \{\neg, \Diamond, \mathsf{E}, \downarrow a.\} \\
([\circ \psi] \varphi)^* &= \downarrow a.\langle a \circ \psi \rangle \varphi^* \text{ for all } \circ \in \{+, -, \circlearrowleft\}, \text{ where } a \text{ is new}
\end{aligned}
$$

It is easy to verify that $\models \varphi \leftrightarrow \varphi^*$ and $\varphi^* \in \mathcal{L}^n$. By axioms in (L2), (2) can be proved by induction on complexity of $\varphi$.

As a conclusion, for each $X \subseteq \{+, -, \circlearrowleft\}$, we have

**Theorem 4.** $\mathsf{C_{LLV}}(X)$ *is sound and strongly complete w.r.t.* $\mathsf{LLV}(X)$.

*Remark 1.* The readers may notice that the expressive power of hybrid logic with downarrow and universal existential operator is as strong as the one of first-order logic, which makes the results in this section less surprising. However, for fragments without the operator $\langle + \rangle$, we can work with the hybrid language $\mathcal{L}(@, \downarrow)$ instead of $\mathcal{L}(\mathsf{E}, \downarrow)$, which leads to stronger results. For example, a sound and complete axiomatization for $\mathsf{LLD}(@, \downarrow)$ is obtained immediately.

## 6   Conclusion

*Summary.* Axiomatization of logic of global and local definable link variations are investigated in this work. For global link variations, the logic $\mathsf{GLV}$ based on hybrid logic $\mathsf{H}(@)$ is introduced and a Hilbert-style calculus $\mathsf{C}_{\mathsf{GLV}}$ is provided. The calculus $\mathsf{C}_{\mathsf{GLV}}$ is shown to be sound and strongly complete with respect to $\mathsf{GLV}$ by constructing a family of canonical models inductively. For local definable link variations, the logic $\mathsf{LLV}$ is introduced. By defining local named dynamic operators and provide recursion axioms, we provide a sound and strongly complete calculus $\mathsf{C}_{\mathsf{LLV}}$ for $\mathsf{LLV}$. For an arbitrarily chosen dynamic fragment $X$, we provide a sound and complete axiomatization for the logic $\mathsf{LLV}(X)$. As a corollary, a sound and complete axiomatization of $\mathsf{LLD}(@, \downarrow)$ is provided, which solves a open problem raised in [12].

*Related Work.* This work is inspired by [8], in which axiomatization for hybrid sabotage logic is provided. The method of completeness proof in this work is a generalization of the one used in [8], and it can be applied to hybrid dynamic logics of other kinds link variations. From the view of graph theory, link variations are special kinds of graph variations and there are many other graph variations, for example, point deletion and point adding. Public announcement logic is one of the logics dealing with point deletion, which was raised in [18]. Local and global public announcement operators have been studied (cf. [4]). Furthermore, the logic of stepwise point deletion is studied in [5], which helps us to understand how the complexity jumps between dynamic epistemic logics of model transformations and logics of randomly chosen graph changes recorded in current memory. Finally, link variation is widely studied in applied logics, for example, social network logics. Nodes in a graph can be viewed as agents, and links can represent the "follow" relation in Twitter or the "friendship" relation in Facebook (cf. [16,17]). Having better understanding of link variations can improve the researches in social network logics.

*Further Directions.* Decidability problem: The logics $\mathsf{GLV}(-)$ and $\mathsf{LLV}(-)$ have been proved to be undecidable (cf. [2,12]). Since $\mathsf{GLV}$ and $\mathsf{LLV}$ extend these logics respectively, they are also undecidable. An immediate technical open problem is to find decidable fragments for these logics.

Different kinds of link variations: In this work, we discuss the dynamic operators link cutting, adding and rotating, which are starting point of the study of more general link variations. What kind of link variation can be axiomatized by the method given in this work? What if there is no hybrid operators? These are all possible directions. In fact, we consider in this work only global undefinable and local definable link variations. Properties of global definable link variations and local undefinable link variations are worth studying.

Applications of logics of link variations: Back to the knowledge graphs, with logics GLV and LLV, we get a better understanding of reasoning in knowledge graphs. Furthermore, in the area of epistemic logic, we could add link variation operators to social network logics and dynamic epistemic logics to make these logics more powerful for reasoning about dynamic situations.

**Acknowledgements.** We thank Johan van Benthem, Fenrong Liu and Dazhu Li for their very helpful suggestions. Thanks for the three anonymous reviewers for their insightful comments for improvements. Thanks also for participants of the conference who asked questions and gave insightful comments. This work is supported by Tsinghua University Initiative Scientific Research Program.

## Appendix: Proof of Lemma 3

*Proof.* Clearly, $f$ is bijective and $f(V^\sigma(x)) = V^{\sigma'}(x)$ for each $x \in \mathsf{Prop} \cup \mathsf{Nom}$. Let $u, v \in W^\Gamma$ and $\sigma' = \sigma * \langle a \circ b \rangle$. Then it suffices to show $\langle u^\sigma, v^\sigma \rangle \in R^\sigma|_{\langle a \circ b \rangle}$ iff $R^{\sigma'} u^{\sigma'} v^{\sigma'}$. One should note that $i \in w^\sigma$ iff $\bar{i} = w^\sigma$ for all $i \in \mathsf{Nom}$ and $w \in W^\Gamma$.

(1) $\circ = +$. Then

$$\langle u^\sigma, v^\sigma \rangle \notin R^\sigma|_{\langle \bar{a} + \bar{b} \rangle} \text{ iff } \langle u^\sigma, v^\sigma \rangle \notin R^\sigma \text{ and } \langle u^\sigma, v^\sigma \rangle \neq \langle \bar{a}, \bar{b} \rangle$$
$$\text{iff } \langle u^\sigma, v^\sigma \rangle \notin R^\sigma \text{ and } \langle a, b \rangle \notin w \times v$$
$$\text{iff } \langle u^{\sigma'}, v^{\sigma'} \rangle \notin R^{\sigma'}$$

(2) $\circ = -$. Then

$$\langle u^\sigma, v^\sigma \rangle \in R^\sigma|_{\langle \bar{a} - \bar{b} \rangle} \text{ iff } \langle u^\sigma, v^\sigma \rangle \in R^\sigma \text{ and } \langle u^\sigma, v^\sigma \rangle \neq \langle \bar{a}, \bar{b} \rangle$$
$$\text{iff } \langle u^\sigma, v^\sigma \rangle \in R^\sigma \text{ and } \langle a, b \rangle \notin w \times v$$
$$\text{iff } \langle u^{\sigma'}, v^{\sigma'} \rangle \in R^{\sigma'}$$

(3) $\circ = \circlearrowleft$. Then we have

$$\langle u^\sigma, v^\sigma \rangle \in R^\sigma|_{\langle \bar{a} \circlearrowleft \bar{b} \rangle} \text{ iff (3.1) } R^\sigma u^\sigma v^\sigma \text{ and } \langle \bar{a}, \bar{b} \rangle \neq \langle u^\sigma, v^\sigma \rangle, \text{ or}$$
$$\text{(3.2) } R^\sigma v^\sigma u^\sigma \text{ and } \langle \bar{a}, \bar{b} \rangle = \langle v^\sigma, u^\sigma \rangle \quad \text{(Proposition 1)}$$
$$\text{iff (3.1') } R^\sigma u^\sigma v^\sigma \text{ and } \langle a, b \rangle \notin u^\sigma \times v^\sigma, \text{ or}$$
$$\text{(3.2') } R^\sigma v^\sigma u^\sigma \text{ and } \langle a, b \rangle \in v^\sigma \times u^\sigma$$
$$\text{iff } \langle u^{\sigma'}, v^{\sigma'} \rangle \in R^{\sigma'}$$

## Appendix: Proof of Lemma 4

*Proof.* The proof of (1) proceeds by induction on the length $n$ of $\sigma$. Let $i, j \in \mathsf{Nom}$ be nominals with $i \in w$ and $j \in v$. Suppose $n = 0$. Then $\varphi \in v$ and $R^\Gamma wv$. Thus $@_i \Diamond j \wedge @_j \varphi \in \Gamma$, which entails $@_i \Diamond \varphi \in \Gamma$ and so $\Diamond \varphi \in w$. Suppose $n > 0$. Then we have three cases:

(1.1) $\sigma = \sigma' * \langle a + b \rangle$. Since $R^\sigma w^\sigma v^\sigma$, either $R^{\sigma'} w^{\sigma'} v^{\sigma'}$ or $\langle a, b \rangle \in w \times v$. Suppose $R^{\sigma'} w^{\sigma'} v^{\sigma'}$. Since $\varphi \in v^\sigma$, we see $\langle a + b \rangle \varphi \in v^{\sigma'}$. By IH[2], $\Diamond \langle a + b \rangle \varphi \in$

---

[2] we write IH for indcution hypothesis.

$w^{\sigma'}$. By axiom (G3,e), $\langle a + b \rangle \Diamond \varphi \in w^{\sigma'}$. Thus $\Diamond \varphi \in w^{\sigma}$. Suppose $\langle a, b \rangle \in w \times v$. Since $\varphi \in v^{\sigma}$, we see $\sigma \varphi \in v$ and so $@_b \sigma \varphi \in w$. By axiom (G3,d), $@_b \langle a + b \rangle \varphi \in w^{\sigma'}$. Note that $a \in w^{\sigma'}$, by axiom (G3,e), $\langle a + b \rangle \Diamond \varphi \in w^{\sigma'}$ and so $\Diamond \varphi \in w^{\sigma}$.

(1.2) $\sigma = \sigma' * \langle a - b \rangle$. Since $R^{\sigma} w^{\sigma} v^{\sigma}$, we see $R^{\sigma'} w^{\sigma'} v^{\sigma'}$ and $\langle a, b \rangle \notin w \times v$. Suppose $a \notin w$. Since $\varphi \in v^{\sigma}$, we see $\langle a - b \rangle \varphi \in v^{\sigma'}$. By IH, $\Diamond \langle a - b \rangle \varphi \in w^{\sigma'}$. Then $\neg a \wedge \Diamond \langle a - b \rangle \varphi \in w^{\sigma'}$. By (G3,f), $\langle a - b \rangle \Diamond \varphi \in w^{\sigma'}$ and so $\Diamond \varphi \in w^{\sigma}$. Suppose $b \notin v^{\sigma'}$. Since $\varphi \in v^{\sigma}$, $\neg b \wedge \langle a - b \rangle \varphi \in v^{\sigma'}$. By IH, $\Diamond (\neg b \wedge \langle a - b \rangle \varphi) \in w^{\sigma'}$. By axiom (G3,f), $\langle a - b \rangle \Diamond \varphi \in w^{\sigma'}$. Thus $\Diamond \varphi \in w^{\sigma}$.

(1.3) $\sigma = \sigma' * \langle a \circlearrowright b \rangle$. Since $R^{\sigma} w^{\sigma} v^{\sigma}$, one of the following cases holds:

(1.3.1) $R^{\sigma'} w^{\sigma'} v^{\sigma'}$ and $\langle a, b \rangle \notin w \times v$. Suppose $b \notin v$. Since $\varphi \in v^{\sigma}$, we see $\neg b \wedge \langle a \circlearrowright b \rangle \varphi \in v^{\sigma'}$. By IH, $\Diamond (\neg b \wedge \langle a \circlearrowright b \rangle \varphi) \in w^{\sigma'}$. By axiom (G3,c) and (G3,h), we have $\langle a \circlearrowright b \rangle \Diamond \varphi \in w^{\sigma'}$ and so $\Diamond \varphi \in w^{\sigma}$. Suppose $b \in v$. Then $a \notin w$. Note that $\langle a \circlearrowright b \rangle \varphi \in v^{\sigma'}$, by IH, $\Diamond \langle a \circlearrowright b \rangle \varphi \in w^{\sigma'}$. Since $a \notin w^{\sigma'}$, we see either $\neg (a \vee b) \wedge \Diamond \langle a \circlearrowright b \rangle \varphi \in w^{\sigma'}$ or $b \wedge \Diamond \langle a \circlearrowright b \rangle \varphi \in w^{\sigma'}$. In both of these cases, one can verify that $\langle a \circlearrowright b \rangle \Diamond \varphi \in w^{\sigma'}$, which entails $\Diamond \varphi \in w^{\sigma}$.

(1.3.2) $R^{\sigma'} v^{\sigma'} w^{\sigma'}$ and $\langle b, a \rangle \in w \times v$. Suppose $a \in w$. Then $w^{\sigma'} = v^{\sigma'}$, which entails $R^{\sigma'} v^{\sigma'} w^{\sigma'}$ and $@_a b \in w^{\sigma'}$. Since $\langle a \circlearrowright b \rangle \varphi \in v^{\sigma'}$, by IH, $\Diamond \langle a \circlearrowright b \rangle \varphi \in w^{\sigma'}$. By axiom (G3,h), we have $\langle a \circlearrowright b \rangle \Diamond \varphi \in w^{\sigma'}$ and so $\Diamond \varphi \in w^{\sigma}$. Suppose $a \notin w$. If $R^{\sigma'} w^{\sigma'} v^{\sigma'}$ holds, then we see $\Diamond \varphi \in w^{\sigma}$ by (3.1). Suppose $v^{\sigma'} \notin R^{\sigma'} (w^{\sigma'})$. Then clearly, $\neg \gamma^{\circlearrowright}_{a,b} \in w^{\sigma'}$. Note that $a \wedge \langle a \circlearrowright b \rangle \varphi \in v^{\sigma'}$, by (G3,h), we have $\Diamond \varphi \in w^{\sigma}$.

The proof of (2) also proceeds by induction on the length $n$ of $\sigma$. Suppose $n = 0$. Since $\Diamond \varphi \in w$, we see $@_i \Diamond \varphi \in \Gamma$ for some $i \in w$. Since $\Gamma$ is pasted, there exists a nominal $j$ such that $@_i \Diamond j \wedge @_j \varphi \in \Gamma$. Let $v = \Delta_j$. Then we see $R^{\Gamma} w v$ and $\varphi \in v$. Suppose $n > 0$. Then we have three cases:

(2.1) $\sigma = \sigma' * \langle a + b \rangle$. Since $\Diamond \varphi \in w^{\sigma}$, $\langle a + b \rangle \Diamond \varphi \in w^{\sigma'}$. By axiom (G3,e), $(a \wedge @_b \langle a + b \rangle \varphi) \vee \Diamond \langle a + b \rangle \varphi \in w^{\sigma'}$. Suppose $a \wedge @_b \langle a + b \rangle \varphi \in w^{\sigma'}$. Then $a \in w^{\sigma'}$ and $@_b \langle a + b \rangle \varphi \in w^{\sigma'}$. Let $v = \Delta_b$. Then $\langle a + b \rangle \varphi \in v^{\sigma'}$, which entails $\varphi \in v^{\sigma}$. Note that $\langle a, b \rangle \in w \times v$, we have $R^{\sigma} w^{\sigma} v^{\sigma}$. Suppose $\Diamond \langle a + b \rangle \varphi \in w^{\sigma'}$. By IH, there is $v \in W^{\Gamma}$ with $R^{\sigma'} w^{\sigma'} v^{\sigma'}$ and $\langle a + b \rangle \varphi \in v^{\sigma'}$. Then $R^{\sigma} w^{\sigma} v^{\sigma}$ and $\varphi \in v^{\sigma}$.

(2.2) $\sigma = \sigma' * \langle a - b \rangle$. Since $\Diamond \varphi \in w^{\sigma}$, we see $\langle a - b \rangle \Diamond \varphi \in w^{\sigma'}$. By axiom (G3,f), $(a \wedge \Diamond (\neg b \wedge \langle a - b \rangle \varphi)) \vee (\neg a \wedge \Diamond \langle a - b \rangle \varphi) \in w^{\sigma'}$. Suppose $a \wedge \Diamond (\neg b \wedge \langle a - b \rangle \varphi) \in w^{\sigma'}$. Then $a \in w^{\sigma'}$ and $\Diamond (\neg b \wedge \langle a - b \rangle \varphi) \in w^{\sigma'}$. By IH, there is $v \in W^{\Gamma}$ such that $R^{\sigma'} w^{\sigma'} v^{\sigma'}$ and $\neg b \wedge \langle a - b \rangle \varphi \in v^{\sigma'}$. Then $b \notin v$ and $\langle a - b \rangle \varphi \in v^{\sigma'}$, which entails $R^{\sigma} w^{\sigma} v^{\sigma}$ and $\varphi \in v^{\sigma}$. Suppose $\neg a \wedge \Diamond \langle a - b \rangle \varphi \in w^{\sigma'}$. Then $a \notin w^{\sigma'}$ and $\Diamond \langle a - b \rangle \varphi \in w^{\sigma'}$. By IH, there is $v \in W^{\Gamma}$ with $R^{\sigma'} w^{\sigma'} v^{\sigma'}$ and $\langle a - b \rangle \varphi \in v^{\sigma'}$. Then $\varphi \in v^{\sigma}$. Since $a \notin w$, $R^{\sigma} w^{\sigma} v^{\sigma}$.

(2.3) $\sigma = \sigma' * \langle a \circlearrowright b \rangle$. Since $\Diamond \varphi \in w^{\sigma}$, $\langle a \circlearrowright b \rangle \Diamond \varphi \in w^{\sigma'}$. By axiom (G3,g), $(\gamma^{\circlearrowright}_{a,b} \wedge \Diamond \langle a \circlearrowright b \rangle \varphi) \vee (\neg \gamma^{\circlearrowright}_{a,b} \wedge \psi) \in w^{\sigma'}$, where $\psi^{\circlearrowright}_{a,b} = (a \wedge \Diamond (\neg b \wedge \langle a \circlearrowright b \rangle \varphi)) \vee (b \wedge (\Diamond \langle a \circlearrowright b \rangle \varphi \vee @_a \langle a \circlearrowright b \rangle \varphi))$. Then there are two cases:

(2.3.1) $\gamma^{\circlearrowright}_{a,b} \wedge \Diamond \langle a \circlearrowright b \rangle \varphi \in w^{\sigma'}$. Then $\Diamond \langle a \circlearrowright b \rangle \varphi \in w^{\sigma'}$. By IH, there is $v \in W^{\Gamma}$ such that $R^{\sigma'} w^{\sigma'} v^{\sigma'}$ and $\langle a \circlearrowright b \rangle \varphi \in v^{\sigma'}$. Then $\varphi \in v^{\sigma'}$. If $\langle a, b \rangle \notin w \times v$, then we obtain $R^{\sigma} w^{\sigma} v^{\sigma}$ immediately. Suppose $\langle a, b \rangle \in w \times v$. Since $\gamma^{\circlearrowright}_{a,b} \in w^{\sigma'}$, we have

$@_a(\neg \Diamond b \vee b) \in w^{\sigma'}$ and so $\neg \Diamond b \vee b \in w^{\sigma'}$. Since $b \in v^{\sigma'}$ and $R^{\sigma'} w^{\sigma'} v^{\sigma'}$, by (1), $\Diamond b \in w^{\sigma'}$. Then $b \in w^{\sigma'}$ and so $w = v$. Since $\langle a, b \rangle \in w \times v$, $R^\sigma w^\sigma v^\sigma$.

(2.3.2) $\neg \gamma^\circlearrowright_{a,b} \wedge \psi^\circlearrowright_{a,b} \in w^{\sigma'}$. Then clearly, exactly one of $a \in w^{\sigma'}$ and $b \in w^{\sigma'}$ holds. Suppose $a \in w^{\sigma'}$. Then $\Diamond(\neg b \wedge \langle a \circlearrowright b \rangle \varphi) \in w^{\sigma'}$. By IH, there is $v \in W^\Gamma$ with $R^{\sigma'} w^{\sigma'} v^{\sigma'}$ and $\neg b \wedge \langle a \circlearrowright b \rangle \varphi \in v^{\sigma'}$. Then $\varphi \in v^\sigma$. Note that $b \notin v$, $R^\sigma w^\sigma v^\sigma$. Suppose $b \in w^{\sigma'}$. Then $\Diamond \langle a \circlearrowright b \rangle \varphi \vee @_a \langle a \circlearrowright b \rangle \varphi \in w^{\sigma'}$. Assume $\Diamond \langle a \circlearrowright b \rangle \varphi \in w^{\sigma'}$. By IH, there is $v \in W^\Gamma$ such that $R^{\sigma'} w^{\sigma'} v^{\sigma'}$ and $\langle a \circlearrowright b \rangle \varphi \in v^{\sigma'}$. Then $\varphi \in v^\sigma$. Note that $a \notin w$, we have $R^\sigma w^\sigma v^\sigma$. Assume $@_a \langle a \circlearrowright b \rangle \varphi \in w^{\sigma'}$. Then $\langle a \circlearrowright b \rangle \varphi \in (\Delta_a)^{\sigma'}$ and so $\varphi \in (\Delta_a)^\sigma$. Since $\langle b, a \rangle \in w \times \Delta_a$, we have $R^\sigma w^\sigma (\Delta_a)^\sigma$.

# References

1. Areces, C., Fervari, R., Hoffmann, G.: Relation-changing modal operators. Log. J. IGPL **23**(4), 601–627 (2015)
2. Areces, C., Fervari, R., Hoffmann, G., Martel, M.: Satisfiability for relation-changing logics. J. Log. Comput. **28**(7), 1443–1470 (2018). https://doi.org/10.1093/logcom/exy022
3. Aucher, G., van Benthem, J., Grossi, D.: Sabotage modal logic: some model and proof theoretic aspects. In: van der Hoek, W., Holliday, W.H., Wang, W. (eds.) LORI 2015. LNCS, vol. 9394, pp. 1–13. Springer, Heidelberg (2015). https://doi.org/10.1007/978-3-662-48561-3_1
4. Belardinelli, F., Ditmarsch, H.V., Hoek, W.: A logic for global and local announcements (2017)
5. Benthem, J.V., Mierzewski, K., Blando, F.Z.: The modal logic of stepwise removal. Rev. Symb. Log. **15**, 36–63 (2020)
6. Benthem, J.: An essay on sabotage and obstruction. In: Hutter, D., Stephan, W. (eds.) Mechanizing Mathematical Reasoning. LNCS (LNAI), vol. 2605, pp. 268–276. Springer, Heidelberg (2005). https://doi.org/10.1007/978-3-540-32254-2_16
7. van Benthem, J.: Logic in games. University of Amsterdam, p. 1 (2014)
8. van Benthem, J., Li, L., Shi, C., Yin, H.: Hybrid sabotage modal logic. J. Log. Comput. **33**, 1216–1242 (2022)
9. van Benthem, J., Mierzewski, K., Blando, F.Z.: The modal logic of stepwise removal. Rev. Symb. Log. **15**(1), 36–63 (2022)
10. Cate, B.: Model theory for extended modal languages. Amsterdam (2005)
11. Hogan, A., Gutierrez, C., Cochcz, M., et al.: Knowledge Graphs. Springer, Cham (2022)
12. Li, D.: Losing connection: the modal logic of definable link deletion. J. Log. Comput. **30**(3), 715–743 (2020)
13. Löding, C.: Sabotage modal logic. In: Foundations of Software Technology and Theoretical Computer Science (2003)
14. Löding, C., Rohde, P.: Model checking and satisfiability for sabotage modal logic. In: Pandya, P.K., Radhakrishnan, J. (eds.) FSTTCS 2003. LNCS, vol. 2914, pp. 302–313. Springer, Heidelberg (2003). https://doi.org/10.1007/978-3-540-24597-1_26
15. Rohde, P.: On games and logics over dynamically changing structures. Ph.D. thesis, Department of Informatics, Technische Hochschule Aachen (RWTH) (2005)
16. Seligman, J., Liu, F., Girard, P.: Logic in the community. In: Banerjee, M., Seth, A. (eds.) ICLA 2011. LNCS (LNAI), vol. 6521, pp. 178–188. Springer, Heidelberg (2011). https://doi.org/10.1007/978-3-642-18026-2_15

17. Seligman, J., Liu, F., Girard, P.: Facebook and the epistemic logic of friendship (2013)
18. Stichting, C., Centrum, M., Baltag, A., Moss, L.S., Solecki, S.: The logic of public announcements. In: Proceedings of the 7th Conference on Theoretical Aspects of Rationality and Knowledge (1998)
19. Stanley, W., Katherine, F.: Dawn: Social Network Analysis: Methods and Applications (Structural Analysis in the Social Sciences). Cambridge University, Cambridge (1994)

# Kleene Algebra of Weighted Programs with Domain

Igor Sedlár$^{(\boxtimes)}$ (ID)

Czech Academy of Sciences, Institute of Computer Science, Prague, Czech Republic
`sedlar@cs.cas.cz`

**Abstract.** Weighted programs were recently introduced by Batz et al. (Proc. ACM Program. Lang. 2022) as a generalization of probabilistic programs which can also represent optimization problems and, in general, programs whose execution traces carry some sort of weight. Batz et al. show that a weighted version of Dijkstra's weakest precondition operator can be used to reason about the competitive ratios of weighted programs. In this paper we study a propositional abstraction of weighted programs with three main contributions. First, we formulate a semantics for weighted programs with the weighted weakest precondition operator based on functions from multimonoids to quantales. Second, we show that the weighted weakest precondition operator corresponds to a generalization of the domain operator known from Kleene algebra with domain, and we study the properties of the generalized domain operator. Third, we formulate a weighted version of Kleene algebra with domain as a framework for reasoning about weighted programs with weakest precondition in an abstract setting.

**Keywords:** Kleene algebra with domain · Kleene algebra with tests · Program semantics · Weakest precondition calculus · Weighted programs

## 1 Introduction

Weighted programs [2] generalize deterministic while programs and probabilistic programs to a framework that can represent optimization problems and, in general, can be used to model programs whose execution traces carry some sort of weight. It is shown in [2] that a weighted version of Dijkstra's weakest precondition operator can be used to reason about the competitive ratios of weighted programs. In [24] a version of Kleene algebra with tests [15] is considered that formalizes reasoning about a propositional abstraction of weighted programs.

In this paper we extend [24] with a formalization of the weighted weakest precondition operator of [2] using a weak version of the domain operator of Kleene algebra with domain [3,4]. In Sect. 2 we introduce our propositional abstraction of weighted programs, We, which can be seen as an extension of Guarded Kleene algebra with tests [26]. In Sect. 3, we propose a semantics for We based

© The Author(s), under exclusive license to Springer Nature Switzerland AG 2024
N. Gierasimczuk and F. R. Velázquez-Quesada (Eds.): DaLí 2023, LNCS 14401, pp. 52–67, 2024.
https://doi.org/10.1007/978-3-031-51777-8_4

on functions from multimonoids to quantales. This semantics is an abstraction of some of the known semantics for Kleene algebra with tests: ordered pairs and guarded strings are replaced by an abstract multimonoid and functions to the two-element set of Boolean truth values (i.e. characteristic functions of sets) are replaced by functions to an arbitrary quantale (complete idempotent semiring). Our functional semantics builds on the work of [6]. In Sect. 4, we show that the weighted weakest precondition operator of [2] can be formalized within our framework using a weak version of the domain operator of Kleene algebra with domain [3,4]. This motivates an extension of We with an additional program operator that corresponds to domain, WeD. In Sect. 5 we define WeKAD, a version of Kleene algebra with tests that formalizes reasoning about WeD. Section 6 briefly discusses related work carried out in [9,10,23,24].

## 2 A Propositional Abstraction of Weighted Programs

The syntax of *deterministic while programs* [1,12] expresses the core of typical imperative programming languages: basic commands are assignments of values (of specific arithmetic expressions) to variables, and complex programs are built up from basic commands and a set of Boolean conditions using the **skip** command, sequential composition (;), conditional branching (**if** ... **then** ... **else** ...) and while loops (**while** ... **do** ...). The state transition semantics of deterministic while programs assigns to each program a partial function on a set of states; intuitively, the function associated with a given program assigns state $s'$ to $s$ iff the execution of the program in $s$ terminates in $s'$, and the function is undefined on $s$ if the execution of the program in $s$ diverges.

*Weighted programs* as defined in [2] are based on the idea that execution traces of programs carry weights, typically taken from some semiring of weights. Weights can be interpreted in probabilistic terms, but also in terms of resource consumption etc. It is argued in [2] that weighted programs constitute a versatile framework for specifying mathematical models (such as optimization problems or probability distributions) in terms of algorithmic representations. Syntactically, weighted programs extend deterministic while programs by operators allowing *non-deterministic branching* and *adding weight* to the current execution trace.

We define a propositional abstraction of weighted programs where basic commands and basic Boolean expressions are represented by variables.

**Definition 1 (Weighted programs).** *Let* P, B *and* E *be disjoint countable sets of variables. The (propositional) language of weighted programs,* $\mathfrak{L}_{We}$, *is defined as follows:*

– *Boolean expressions:*   $B, C ::= \mathsf{b} \in \mathsf{B} \mid \top \mid \bot \mid \neg B \mid B \wedge C \mid B \vee C$
– *Weight expressions:*   $E, F ::= \mathsf{e} \in \mathsf{E} \mid 1 \mid 0 \mid E \cdot F \mid E + F$
– *Programs:*
  $P, Q ::= \mathsf{p} \in \mathsf{P} \mid P; Q \mid \mathsf{if}\ B\ \mathsf{then}\ P\ \mathsf{else}\ Q \mid \mathsf{while}\ B\ \mathsf{do}\ P \mid P \oplus Q \mid \odot E$

*We define* PEB $=$ P $\cup$ E $\cup$ B. *The set of all programs (Boolean expressions, weight expressions) is denoted as* Pr *(Bt, Wt). We define* Exp $=$ Pr $\cup$ Bt $\cup$ Wt.

Boolean expressions are formulas of the language of classical propositional logic. Weight expressions represent weights taken from some abstract semiring of weights (see the semantics specified in Sect. 3). Programs extend the syntax of deterministic while programs with non-deterministic branching $\oplus$ and commands of the form $\odot E$ read as "add weight (corresponding to) $E$ to the current execution trace of the program". We will write $P \odot E$ instead of $P; \odot E$.

*Example 1.* Consider the weighted program

$$\textbf{while } b \textbf{ do } (p \odot e) \tag{1}$$

This is an ordinary "while b do p" loop, but now with addition of the weight (corresponding to) e after each iteration of the basic command p.

*Example 2.* If e represents a value $x \in [0,1]$ and $\bar{e}$ represents $1 - x$, then the program

$$(P \odot e) \oplus (Q \odot \bar{e}) \tag{2}$$

represents a sort of probabilistic branching: execute $P$ with probability $x$ and $Q$ with probability $1 - x$.

Weighted programs as defined here can be seen as an extension of Guarded Kleene Algebra with Tests (GKAT); see [26]. To see this, note that we can define **skip** $:= \odot 1$, **abort** $:= \odot 0$, and **assert** $B := \textbf{if } B \textbf{ then skip else abort}$. In fact, weighted programs are related to ProbGKAT, the recently introduced probabilistic extension of GKAT [23]; see Sect. 6.

## 3   Semantics for Weighted Programs

It is natural to generalize the state transition semantics for deterministic while programs to a semantics for weighted programs where the interpretation of a program $P$ is a binary $S$-weighted relation on a set of states $U$ for some semiring $S$, that is, a mapping $\tilde{V}(P) : (U \times U) \to S$. Intuitively, the value of $\tilde{V}(P)$ at $(s, s')$ is the minimal weight of an execution trace of $P$ that starts in $s$ and terminates in $s'$. In this section, we formulate a variant of such a semantics that is more general in one sense and more specific in another sense. In particular, we replace $U \times U$ by a *multimonoid* [6,17] and our weight algebras will be *unital quantales* instead of general semirings.

### 3.1   Semirings and Quantales

We assume familiarity with the notion of a (complete, idempotent) *semiring*. We will usually refer to algebras using the name of their universe, so a semiring $(S, +, \cdot, 0, 1)$ will in general be referred to as $S$.

*Example 3.* We give four examples of complete idempotent semirings. (1) The *Boolean semiring* is $(2, \vee, \wedge, 0, 1)$, where $2 = \{0, 1\}$.

(2) The *tropical semiring* over extended natural numbers, well known especially from shortest path algorithms, is $(\mathbb{N}^\infty, \min, +, \infty, 0^\mathbb{N})$, where $\mathbb{N}^\infty = \mathbb{N} \cup \{\infty\}$, $\infty$ is the top element of the semiring with respect to the partial order induced by semiring addition min, and natural number addition is seen as semiring multiplication with $0^\mathbb{N}$ (natural number zero) as neutral element.

(3) The *Łukasiewicz semiring* is $L = ([0, 1], \max, \&, 0, 1)$ where $[0, 1]$ is the real unit interval and $\&$ is the Łukasiewicz t-norm $x \& y = \max\{0, x + y - 1\}$. The Łukasiewicz semiring is well known from fuzzy logic.

(4) The *Viterbi semiring* (or the product semiring), well known from probabilistic models, is $\Pi = ([0, 1], \max, \times, 0, 1)$ where $\times$ is multiplication in the real unit interval.

A semiring can be seen as an abstract representation of *weights* (or costs). In particular, semiring multiplication corresponds to weight addition ($x \cdot y$ is the result of adding weight $y$ to weight $x$) and so the multiplicative identity 1 corresponds to "zero weight" (since $x \cdot 1 = x = 1 \cdot x$). The annihilator element 0 corresponds to "absolute weight" (since $0 \cdot x = 0 = x \cdot 0$). Semiring addition reflects, in a sense, ordering of weights. In idempotent semirings, $x + y$ can be seen as the minimal weight among $\{x, y\}$.

Idempotent semirings (also called dioids) are especially fitting as a model of weights in settings where one considers a set of objects (such as execution traces of a program), each associated with a weight, and wants to select the "optimal choice" among the objects. Often the set of objects to choose from is infinite, however, and so one is naturally led to considering *complete idempotent semirings* of weights. Such semirings are a special case of *quantales* [18,20].

**Definition 2 (Quantale).** *A quantale is a structure $(Q, \leq, \cdot)$ where $(Q, \leq)$ is a complete lattice with supremum $\bigvee$ and infimum $\bigwedge$, and $\cdot$ is a binary operation that distributes into joins:*

$$x \cdot \left( \bigvee_{i \in I} y_i \right) = \bigvee_{i \in I} (x \cdot y_i) \qquad \left( \bigvee_{i \in I} y_i \right) \cdot x = \bigvee_{i \in I} (y_i \cdot x).$$

*A quantale is* unital *iff there is a unique $1 \in Q$ such that $x \cdot 1 = x = 1 \cdot x$.*

Complete idempotent semirings are unital quantales.[1] In what follows, we will refer to unital quantales simply as quantales. Each quantale gives rise to a residuated lattice [8]. Every quantale is a (*-continuous) *Kleene algebra* [13, 14] where $x^* = \bigvee_{n \in \mathbb{N}} x^n$. All semirings mentioned in Example 3 are complete idempotent semirings, hence quantales.

---

[1] If $(S, +, 0)$ is a complete idempotent commutative monoid, then it is a complete join-semilattice. Every complete join-semilattice gives rise to a complete lattice. Note that $0 = \bigvee \emptyset$.

## 3.2  Multimonoids

The set $U \times U$ over some $U$ gives rise to a monoid-like structure where the multiplication operation is partial. A suitable generalization of this example is given by the notion of a multimonoid [17].

**Definition 3 (Multimonoid).** *A* multimonoid *is a structure* $(M, \otimes, I)$ *such that $M$ is a non-empty set,* $\otimes : M \times M \to 2^M$ *is associative and $I \subseteq M$ such that, for all $K \subseteq M$, $K \otimes I = K = I \otimes K$ (assuming the lifting of $\otimes$ to subsets of $M$ defined by $K \otimes L = \{z \mid \exists x, y \in M(x \in K \& y \in L \& z \in x \otimes y)\}$).*
    *A multimonoid is* local *iff, for all $x, y, z, u$, if $u \in x \otimes y$ and $y \otimes z \neq \emptyset$, then $u \otimes z \neq \emptyset$.*

The operation $\otimes$ in multimonoids is nothing but a ternary relation on $M$. In multimonoids, every $x \in M$ has a unique *left identity* in $I$ (that is, an element $e \in I$ such that $e \otimes x = \{x\}$) and similarly a unique *right identity* in $I$ (see Appendix A.1). Local multimonoids will be important in Sect. 4 and we postpone a more thorough discussion of their properties until then.

*Remark 1.* Multimonoids are related to relational frames for relevant and sub-structural logics [19,21]. These are frames of the form $(U, \leq, R, N)$ where $R$ is a ternary relation on the partially ordered set $U$, and $N$ is a subset of $U$ that is closed under $\leq$ and (i) for all $x \in U$ there is $y \in N$ such that $Ryxx$, and (ii) for all $x, y, z \in U$, if $x \in N$ and $Rxyz$, then $y \leq z$. In particular, a multimonoid is a frame where $\leq$ is the discrete order, $R$ is fully associative[2] and $N$ in addition satisfies (iii) for all $x \in U$ there is $y \in N$ such that $Rxyx$, and (iv) for all $x, y, z \in U$, if $x \in N$ and $Ryxz$, then $y \leq z$.

*Example 4.* We give two examples of local multimonoids. Many other examples (of local and non-local multimonoids) are provided in [6,17].

(1) The relational multimonoid over $U$ is $(U \times U, \otimes, \mathrm{id}_U)$ where $(x, y) \otimes (y', z)$ is $\{(x, z)\}$ in case $y = y'$ and $\emptyset$ otherwise.
(2) Take a finite subset $\mathsf{A}$ of the set of Boolean variables $\mathsf{B}$. Assume that $\mathsf{A}$ is ordered in some arbitrary but fixed way as $\mathsf{a}_1, \ldots, \mathsf{a}_n$ for $\mathsf{a}_i \in \mathsf{A}$. An *atom* (over $\mathsf{A}$) is a sequence $\mathsf{c}_1, \ldots, \mathsf{c}_n$ where each $\mathsf{c}_i$ is either $\mathsf{a}_i$ or $\bar{\mathsf{a}}_i$. A *guarded string* over $\mathsf{A}$ and $\mathsf{P}$ [16] is a finite sequence $A_1 p_1 \ldots A_{n-1} p_{n-1} A_n$, where each $A_i$ is an atom over $\mathsf{A}$ and each $p_i \in \mathsf{P}$. Let $G_{\mathsf{A},\mathsf{P}}$ be the set of all guarded strings over $\mathsf{A}$ and $\mathsf{P}$. The *coalesced product* of two guarded strings is defined as follows:

$$wA \diamond Bu = \begin{cases} wAu & \text{if } A = B \\ \text{undefined} & \text{otherwise.} \end{cases}$$

The coalesced product can be naturally expressed as a ternary relation on guarded strings and so $G_{\mathsf{A},\mathsf{P}}$ is an example of a multimonoid where $I$ is the set of all atoms.

---

[2] There is $w$ such that $Rxyw$ and $Rwzv$ iff there is $u$ such that $Ryzu$ and $Rxuv$.

Let $Q^M$ be the set of all functions from a multimonoid $M$ to a quantale $Q$.

**Lemma 1.** $Q^M$ *is a quantale.*

*Proof.* See [6], Theorem 4.5. The partial order and monoid multiplication in $Q^M$ are defined point-wise in the obvious fashion; the multiplicative identity in $Q^M$ is the function that assigns the multiplicative identity of $Q$ to $x \in I$ and the annihilator element of $Q$ to all other elements of $M$. ☐

### 3.3 Functional Semantics

A function $f \in Q^M$ is *diagonal* iff $f(x) = 0$ for $x \notin I$, *crisp* if $f(x) \in \{1, 0\}$ and *diagonally constant* (di-constant) if $f(x) = f(y)$ for all $x, y \in I$. A *predicate* is a crisp diagonal function $f \in Q^M$; a diagonal function may also be called a *weighted predicate.*

**Definition 4 (Model).** *Let $M$ be a multimonoid and $Q$ a quantale. An $Q^M$-model is a function $V : \mathsf{PEB} \to Q^M$ such that*

- $V(\mathsf{p})$ *is crisp for all* $\mathsf{p} \in \mathsf{P}$*;*
- $V(\mathsf{b})$ *is a predicate for all* $\mathsf{b} \in \mathsf{B}$*; and*
- $V(\mathsf{e})$ *is diagonal and diagonally constant for all* $\mathsf{e} \in \mathsf{E}$*.*

*A $Q^M$-model is* local *iff $M$ is a local multimonoid.*

**Definition 5 (Interpretation).** *Given a $Q^M$-model $V$, we define the function $\tilde{V} : Exp \to Q^M$ as follows:*

- $\tilde{V}(\xi) = V(\xi)$ *for* $\xi \in \mathsf{PEB}$*;*
- $\tilde{V}(B)$ *for Boolean expressions and $\tilde{V}(E)$ for weight expressions are obtained using the obvious lifting of Boolean operations to crisp functions and semiring operations to all functions, respectively;*
- $\tilde{V}(P; Q) = \tilde{V}(P) \cdot \tilde{V}(Q)$*;*
- $\tilde{V}(\mathbf{if}\ B\ \mathbf{then}\ P\ \mathbf{else}\ Q) = (\tilde{V}(B) \cdot \tilde{V}(P)) + (\tilde{V}(\neg B) \cdot \tilde{V}(Q))$*;*
- $\tilde{V}(\mathbf{while}\ B\ \mathbf{do}\ P) = \left(\tilde{V}(B) \cdot \tilde{V}(P)\right)^* \cdot \tilde{V}(\neg B)$*;*
- $\tilde{V}(P \oplus Q) = \tilde{V}(P) + \tilde{V}(Q)$*;*
- $\tilde{V}(\odot E) = \tilde{V}(E)$*.*

## 4 Weighted Predicate Transformers and Domain

### 4.1 Predicate Transformers and Weightings

Dijkstra's *predicate transformer* semantics for an extension of while programs [5] assigns to each program $P$ a function $\mathsf{wp}[\![P]\!] : 2^U \to 2^U$ such that $\mathsf{wp}[\![P]\!](Y)$ is the set of all states $x$ such that each computation of $P$ starting in $x$ terminates in some state in $Y$. The predicate (set of states) $\mathsf{wp}[\![P]\!](Y)$ is known as the *weakest precondition* of $P$ with respect to postcondition $Y$. Dijkstra develops a calculus that allows to compute $\mathsf{wp}[\![P]\!](Y)$ for each $P$ and $Y$. A variant of the weakest precondition operator is the weakest *liberal* precondition operator where $\mathsf{wlp}[\![P]\!](Y)$ is the set of states $x$ such that each terminating computation of $P$ starting in $x$ terminates in a state in $Y$ (not all computations of $P$ starting in $x$ must terminate). Note that $\mathsf{wlp}[\![P]\!]$ is reminiscent of the box operator of Propositional Dynamic Logic; see [7]. The natural dual of the weakest liberal precondition operator is the weakest *"angelic"* precondition operator such that $\mathsf{wap}[\![P]\!](Y)$ is the set of states $x$ such that *some* computation of $P$ starting in $x$ terminates in a state in $Y$. Note that $\mathsf{wap}[\![P]\!]$ is reminiscent of the diamond operator of Propositional Dynamic Logic.

Batz et al. [2] generalize the notion of weakest angelic precondition as follows. First, the notion of a predicate (essentially, a function in $\{1,0\}^X$ or a crisp diagonal function in $\{1,0\}^{X\times X}$) is generalized to the notion of a *weighted predicate* (called a "weighting" by Batz et al.), that is, a function in $S^U$ or, equivalently, a diagonal function in $S^{U\times U}$. Second, a *weighted predicate transformer semantics* is defined for each program $P$ where $\mathsf{wap}[\![P]\!] : S^U \to S^U$ operates as follows when $\mathsf{wap}[\![P]\!](w)$ is applied to a state $s \in U$: first, the function takes all terminating execution traces $\tau \in T$ of $P$ starting in $x$ and computes the accumulated weights of the traces, obtaining a set of values $\{t_\tau \mid \tau \in T\}$ for each trace $\tau \in T$; second, the function adds to the accumulated weight of each trace $\tau$ the value of $w$ at the final state $y_\tau$ of the trace, thus obtaining $\{t_\tau \cdot w(y_\tau) \mid \tau \in T\}$; and, third, $\mathsf{wap}[\![P]\!](w)(x)$ returns

$$\sum_{\tau \in T} \{t_\tau \cdot w(y_\tau)\} .$$

Hence, informally, $\mathsf{wap}[\![P]\!](1)(x)$ is the weight of the "least expensive" execution trace of $P$ starting in $x$. Batz et al. [2] develop a weakest angelic preweighting calculus that allows to compute $\mathsf{wap}[\![P]\!]$ for each weighted program $P$, and they apply the calculus to reasoning about competitive ratios of weighted programs.

## 4.2   Domain

Recall that in a multimonoid $M$, each element has a unique left and right identity in $I$. That is, for each $x \in M$ there is a unique $y \in I$ such that $y \otimes x = \{x\}$, and a unique $z \in I$ such that $x \otimes z = \{x\}$. Let us denote the unique left identity of $x$ as $\mathsf{s}(x)$ (for "source") and the unique right identity as $\mathsf{t}(x)$ (for "target"); see [6].

We can formalize the $\mathsf{wap}$ operator directly in $Q^M$-models for weighted programs as follows. Let $Di(Q^M)$ be the set of all diagonal functions (weighted predicates) in $Q^M$.

**Definition 6 (Wap).** *Given a $Q^M$-model $V$, we define $W : Pr \to (Di(Q^M) \to Q^M)$ by stipulating that, for all $f \in Di(Q^M)$ and $x \in M$,*

$$W(P)(f)(x) = \bigvee_{x=\mathsf{s}(y)} \left( \tilde{V}(P)(y) \cdot f(\mathsf{t}(y)) \right) . \tag{3}$$

It is easily seen that $W(P)(f) \in Di(Q^M)$ (this follows from $\mathsf{s}(y) \in I$). The restriction of the second argument of $W$ to $Di(Q^M)$ is not essential.

It can be shown that the $W$ operator can be analysed in terms of a generalisation of the domain operator known from Kleene algebra with domain [3,4] (the operator also appears in Dynamic Predicate Logic [11]). This particular generalisation of the domain operator has been considered in [6].

**Definition 7 (Domain).** *We define* $\mathsf{d} : Q^M \to Q^M$ *as follows:*

$$\mathsf{d}(f)(x) = \bigvee_{x=\mathsf{s}(y)} f(y) \, . \tag{4}$$

It is easily seen that $\mathsf{d}(f) \in Di(Q^M)$ for all $f \in Q^M$. Proposition 1 specifies some useful properties of d. Proposition 2 points out some familiar properties of the domain operator in Kleene algebra with domain that fail in the present setting.

**Proposition 1.** *The following hold for all* $f, g \in Q^M$ *and* $h \in Di(Q^M)$:

1. $\mathsf{d}(h) = h$
2. $\mathsf{d}(f \cdot g) \leq \mathsf{d}(f \cdot \mathsf{d}(g))$
3. $\mathsf{d}(\mathsf{d}(f) \cdot g) = \mathsf{d}(f) \cdot \mathsf{d}(g)$
4. $\mathsf{d}\left(\bigvee_{i \in I} f_i\right) = \bigvee_{i \in I} \mathsf{d}(f_i)$

*If $M$ is a local multimonoid, then:*

5. $\mathsf{d}(f \cdot \mathsf{d}(g)) \leq \mathsf{d}(f \cdot g)$

*Proof.* See Appendix A.2.

**Proposition 2.** *The following do not hold for all* $f, g \in Q^M$:

1. $\mathsf{d}(f) \leq 1$
2. $f \leq \mathsf{d}(f) \cdot f$

*Proof.* See Appendix A.3.

One of the interesting questions for future work is to characterize the properties of $f \in Q^M$ in terms of the domain axioms $f$ satisfies.

The main observation of this subsection is that d can be used to define the $W$ operator. The salient fact is expressed in the following proposition.

**Proposition 3.** *For all* $f \in Q^M$, $g \in Di(Q^M)$ *and* $x \in M$:

$$\mathsf{d}(f \cdot g)(x) = \bigvee_{x=\mathsf{s}(y)} (f(y) \cdot g(\mathsf{t}(y))) \, . \tag{5}$$

*Hence,* $W(P)(g) = \mathsf{d}(\tilde{V}(P) \cdot g)$.

*Proof.* See Appendix A.4.

We use Proposition 3 to show that the $W$ operator has the properties of the wap operator established in [2] (Table 2 on p. 14). However, we need to restrict the claim to local models.

**Theorem 1.** *Let $V$ be a local $Q^M$-model and let $f \in Di(Q^M)$. The following claims hold for arbitrary $E, B, P$ and $Q$ ("lfp" means "least fixed point"):*

1. $W(P;Q)(f) = W(P)(W(Q)(f))$;
2. $W(\text{if } B \text{ then } P \text{ else } Q)(f) = \tilde{V}(B) \cdot (W(P)(f)) + \tilde{V}(\neg B) \cdot (W(Q)(f))$;
3. $W(\text{while } B \text{ do } P)(f) = \text{lfp}.\xi \left( \tilde{V}(\neg B)(f) + \tilde{V}(B)W(P)(\xi) \right)$;
4. $W(P \oplus Q)(f) = W(P)(f) + W(Q)(f)$;
5. $W(\odot E)(f) = \tilde{V}(E) \cdot f$.

*Proof.* We show that the properties of $W$ stated in the theorem follow from certain facts about arbitrary *-continuous Kleene algebras with tests expanded with a unary operator d satisfying the properties of Proposition 1.[3] By Lemma 1 and Proposition 1, $Q^M$ is such a Kleene algebra (predicates obviously form a Boolean algebra).

(1.) It is sufficient to show that $\mathsf{d}(pqr) = \mathsf{d}(pd(qr))$. This holds by Proposition 1(2., 5.).

(2.) It is sufficient to show that $\mathsf{d}((bp + \bar{b}q)r) = b\mathsf{d}(pr) + \bar{b}\mathsf{d}(qr)$. This is established as follows:

$$\mathsf{d}((bp + \bar{b}q)r) = \mathsf{d}(bpr + \bar{b}qr)$$
$$= \mathsf{d}(bpr) + \mathsf{d}(\bar{b}qr) \qquad \text{Prop. 1(4.)}$$
$$= \mathsf{d}(\mathsf{d}(b)pr) + \mathsf{d}(\mathsf{d}(\bar{b})qr) \qquad \text{Prop. 1(1.)}$$
$$= b\mathsf{d}(pr) + \bar{b}\mathsf{d}(qr) \qquad \text{Prop. 1(1., 3.)}$$

(3.) It is sufficient to show that if $q = \mathsf{d}(q)$, then $\mathsf{d}((bp)^*\bar{b}q) = \alpha$ is the least prefixed point of the function $\phi : e \mapsto (\bar{b}q + \mathsf{bd}(pe))$. First, we show that $\phi(\alpha) \leq \alpha$. Since $1 \leq (bp)^*$, we have $\mathsf{d}(\bar{b}q) \leq \alpha$. However $\mathsf{d}(\bar{b}q) = \bar{b}q$ by Prop. 1(1., 3.) and the assumption $q = \mathsf{d}(q)$. Hence, $\bar{b}q \leq \alpha$. Next, $\mathsf{bd}(p\alpha) = \mathsf{d}(bp(bp)^*\bar{b}q)$ by Prop. 1(1.,2.,3.,5.) and so $\mathsf{bd}(p\alpha) \leq \alpha$ since $rr^* \leq r^*$ in Kleene algebra. Hence, $\bar{b}q + \mathsf{bd}(p\alpha) = \phi(\alpha) \leq \alpha$.

Now we show that $\alpha$ is the least pre-fixed point of $\phi$. In fact, it is sufficient to show that the following holds in each KAT satisfying our assumptions (for arbitrary $p, q, r$):

$$\mathsf{d}(q + pr) \leq \mathsf{d}(r) \implies \mathsf{d}(p^*q) \leq \mathsf{d}(r)$$

In a *-continuous KAT, $p^*q = \bigvee_{n \in \mathbb{N}}(p^n q)$. Since d is assumed to be completely additive, $\mathsf{d}(p^*q) = \bigvee_{n \in \mathbb{N}} \mathsf{d}(p^n q)$. We prove that $\mathsf{d}(q + pr) \leq \mathsf{d}(r)$ implies $\mathsf{d}(p^n q) \leq \mathsf{d}(r)$ for all $n \in \mathbb{N}$. We will not refer to individual claims of Prop. 1 that justify

---

[3] In particular, the KAT has to satisfy the properties that result from the ones in Proposition 1 by replacing $f, g$ with arbitrary elements of the KAT and $h$ with an arbitrary element of the Boolean algebra of tests.

our steps any longer. Base case: $d(q + pr) \le d(r)$ entails $d(q) + d(pr) \le d(r)$ and so $p^0 d(q) \le d(r)$; but this means that $d(p^0 q) \le d(r)$ since $p^0 = 1$. Induction step: we prove that if $d(q + pr) \le d(r) \Rightarrow d(p^n q) \le d(r)$, then $d(q + pr) \le d(r) \Rightarrow d(p^{n+1}q) \le d(r)$. We assume $d(q + pr) \le d(r)$ and we reason as follows:

$$d(p^n q) \le d(r)$$
$$pd(p^n q) \le pd(r)$$
$$d(pd(p^n q)) \le d(\le pd(r))$$
$$d(p^{n+1}q) \le d(pr) \le d(r)$$

(The last inequation holds by the assumption $d(q + pr) \le d(r)$.)

(4.) is a trivial consequence of (finite) additivity of d: $d(p + q) = d(p) + d(q)$.
(5.) It is sufficient to notice that if $p = d(p)$ and $q = d(q)$, then $d(pq) = pq$.     □

### 4.3  Weighted Programs with Domain

The above result suggests that the weakest angelic preweighting operator could be integrated into weighted programs by introducing an additional unary program operator $\Diamond$ corresponding to d via $\tilde{V}(\Diamond P) = d(\tilde{V}(P))$. We call this extension *weighted programs with domain*, WeD. The language of WeD does not have specific variables for weighted predicates, but note that $\tilde{V}(\Diamond P)$ is a weighted predicate for each program $P$. Therefore, we can consider $\{\Diamond P \mid P \in Pr\}$ as the set of expressions denoting weighted predicates. This is similar to how Boolean predicates are expressed in one-sorted Kleene algebra with domain [4].

## 5  Weighted Kleene Algebra with Domain

In this section we abstract away from the multimonoid semantics of WeD and we define a suitable class of Kleene algebras. These algebras extend Kleene algebras with weights and tests [24] (by adding the domain operator) which in turn extend Kleene algebras with tests [15,16]. This move is quite natural since we have already benefited from the fact that $Q^M$ is a Kleene algebra.

**Definition 8 (WeKAD language).** *The language of weighted Kleene algebra with domain ($\mathfrak{L}_{\mathsf{WeKAD}}$) contains two sorts of terms, namely, Boolean terms and programs:*

- *Boolean terms:*  $b, c ::= \mathsf{b} \in \mathsf{B} \mid 1 \mid 0 \mid \bar{b} \mid b \cdot c \mid b + c$
- *Programs:*  $p, q ::= \mathsf{p} \in \mathsf{P} \mid \mathsf{e} \in \mathsf{E} \mid b \mid p \cdot q \mid p + q \mid p^* \mid d(p)$

The language $\mathfrak{L}_{\mathsf{WeKAD}}$ extends the language of Kleene algebra with tests $\mathfrak{L}_{\mathsf{KAT}}$ with the weight variables $\mathsf{e} \in \mathsf{E}$ and the domain operator d; it also extends the (two-sorted) language of Kleene algebra with domain $\mathfrak{L}_{\mathsf{KAD}}$ [3] with weight variables. In fact, however, $\mathfrak{L}_{\mathsf{WeKAD}}$ can be seen as a version of $\mathfrak{L}_{\mathsf{KAD}}$ where $\mathsf{P} \cup \mathsf{E}$ is seen as the set of program variables.

**Definition 9 (WeKAD).** *A weighted Kleene algebra with domain is an algebra of the form*

$$(K, B, Q, +, \cdot, {}^{*}, {}^{-}, \mathsf{d}, 0, 1)$$

*such that (i)* $(K, B, +, \cdot, {}^{*}, {}^{-}, 0, 1)$ *is a Kleene algebra with tests, (ii)* $\{0, 1\} \subseteq Q \subseteq K$ *such that* $(Q, +, \cdot, 1)$ *is a unital quantale, and (iii)* $\mathsf{d} : K \to K$ *such that* $(i \in B \cup Q)$

$$\mathsf{d}(i) = i \tag{6}$$
$$\mathsf{d}(p + q) = \mathsf{d}(p) + \mathsf{d}(q) \tag{7}$$
$$\mathsf{d}(p \cdot \mathsf{d}(q)) = \mathsf{d}(p \cdot q) \tag{8}$$
$$\mathsf{d}(\mathsf{d}(p) \cdot q) = \mathsf{d}(p) \cdot \mathsf{d}(q) \tag{9}$$
$$\mathsf{d}(q + pr) \leq \mathsf{d}(r) \ \Rightarrow \ \mathsf{d}(p^{*}q) \leq \mathsf{d}(r) \tag{10}$$

*A WeKAD valuation is any function* $v : \mathsf{PEB} \to K$ *such that* $v(\mathsf{e}) \in Q$ *and* $v(\mathsf{b}) \in B$. *Validity of equations is defined as expected.*

We note that for each local $M$, $Q^{M}$ with $\mathsf{d}$ defined by (4) forms a WeKAD: define $B$ as the set of predicates and $Q$ as the set of weighted predicates; the first four domain properties were established in Proposition 1, and the final one was established in the proof of Theorem 1. Note that $\mathsf{d}$ does not need to be completely additive in WeKADs; we assume axiom (10) instead.

We did not define $\mathsf{d}$ as a function $K \to Q$ on purpose. This reflects our "intended reading" of $Q$ as the quantale of weights, not the quantale of all weighted predicates. (This reading is somewhat at odds with the remark concerning $Q^{M}$ in the previous paragraph; see also Problem 3 below.) Weighted predicates can be seen as elements $x \in K$ such that $\mathsf{d}(x) = x$.

At this point, we were able only to scratch the surface of WeKAD. There is a number of interesting questions we have to leave for future research:

**Problem 1:** What is the complexity of the equational theory of WeKAD? The equational theory of (one-sorted) KAD is EXPTIME-complete [25], and we expect the same for WeKAD.

**Problem 2:** Is the equational theory of WeKAD identical to the equational theory of some "concrete" subclass of WeKAD, for example the class of algebras based on $Q^{M}$ for local $M$ where $\mathsf{d}$ defined by (4)?

**Problem 3:** WeKAD does not explicitly distinguish between elements of $Q$ that represent weighted predicates in general and elements that represent constant predicates (diagonally constant diagonal functions in the multimonoid setting that correspond to elements of the weight quantale). For instance, in $Q^{M}$ nothing prevents the valuation $v$ from assigning a diagonal function $f \in Q^{M}$ that is not diagonally constant to a weight variable. Similarly, nothing prevents the valuation $v$ from assigning a non-crisp function $f \in Q^{M}$ to a program variable. Can these distinctions be expressed by (quasi)equations in the language of WeKAD?[4]

**Problem 4:** It is natural to consider "axiomatic extensions" of WeKAD. For instance, one can add axioms that make $Q$ an *MV-algebra* (related to the Lukasiewicz quantale $L$) or a *product algebra* (related to $\Pi$); see [8].[5] It would be interesting to look at these extensions in general.

## 6  Related Work

In [24], a form of KAT for (the propositional version of) weighted programs is introduced. This is, essentially, WeKAD minus the domain operator.[6] Semantics of the sort given here is discussed, but it is formulated in terms of so-called *partial semigroups with identity* which are a special case of local multimonoids.

Versions of KAT where tests do not form a Boolean algebra are studied in [9,10]. The motivation is to formalize reasoning about programs, such as fuzzy controllers, where conditions are not Boolean but may take a value in a commutative and integral *residuated lattice*.[7] There is a close connection to our approach since, as we noted, every quantale gives rise to a residuated lattice, which however does not need to be commutative nor integral. Boolean tests are not considered in [9,10], but this is due to the difference in motivation.

ProbGKAT, a probabilistic extension of GKAT, is studied in [23]. In particular, ProbGKAT adds to GKAT return variables (which we do not consider), probabilistic branching $p \oplus_e q$ ("do $p$ with probability $e$ and $q$ with probability $\bar{e} = 1 - e$"), and probabilistic loops $p^{[e]}$ which, at each stage of the loop, execute $p$ with probability $e$ and terminate with probability $1 - e$. ProbGKAT can be seen as using the product quantale $\Pi$ expanded with bounded subtraction. ProbGKAT is thus related to a version of We over this quantale. Różowski and Silva [22] have recently introduced probabilistic regular expressions, PRE, a fragment of ProbGKAT using only probabilistic branching, sequential composition and probabilistic loops. PRE is related to a version of the framework from [24] using the product quantale $\Pi$. The exact nature of the relations between We and the framework from [24] on one hand and PRE and ProbGKAT on the other hand will be determined in future work.

## 7  Conclusion

We studied a propositional abstraction of weighted programs [2] with weighted weakest precondition. In particular, (i) we defined a semantics for weighted programs based on functions from multimonoids to quantales [6]; (ii) we have shown

---

[4] E.g. $d(p)p = p$ is a natural candidate for a "definition" of crisp elements but it does not seem to be adequate.

[5] These would in fact require $^-$ to be defined on $Q$ as well.

[6] In addition, weighted predicates are represented by a semiring, not a quantale.

[7] In these residuated lattices, also known as $FL_{ew}$-algebras, multiplication is commutative and the multiplicative identity is the top element; see [8].

that weighted weakest precondition can be formalized using a weak version of the domain operator of Kleene algebra with domain (Theorem 1); and (iii) we outlined WeKAD, a weighted version of Kleene algebra with domain that is suitable for reasoning about weighted programs. In many respects, the present paper just sets the stage for future developments and technical results.

**Acknowledgement.** This work was supported by the grant 22-16111S of the Czech Science Foundation.

# A      Technical appendix

## A.1      Identities in multimonoids

**Lemma 2.** *Let $M$ be a multimonoid. For all $x, y \in M$ and $i, j \in I$:*

*1. $y \in (x \otimes i) \cup (i \otimes x)$ only if $x = y$.*    *4. $i \otimes x = \{x\} = j \otimes x$ only if $i = j$.*
*2. $x \in i \otimes j$ only if $i = j$.*
*3. $i \otimes i = \{i\}$.*

*5. $x \otimes i = \{x\} = x \otimes j$ only if $i = j$.*

*Proof.* The proof is based on some of the arguments in [6]. (1.) This holds since $\{x\} \otimes I = \{x\} = I \otimes \{x\}$ by definition. (2.) If $x \in i \otimes j$, then $x = i$ and $x = j$ by the previous item. (3.) By definition, $i \otimes I = \{i\}$; hence there is $j \in I$ such that $i \in i \otimes j$. By the previous item, $i \in i \otimes i$. If $x \in i \otimes i$, then $x = i$ by the first item. (4.) If the assumption holds, then $\emptyset \neq i \otimes x = i \otimes (j \otimes x) = (i \otimes j) \otimes x$. It follows that $y \in i \otimes j$ for some $y$, and so $i = j$ by the second item. (5.) is similar.    □

## A.2      Proof of Proposition 1

Before we give the proof, we state a useful lemma of [6] (the following is an excerpt of their Lemmas 3.1 and 3.3). In the lemma, we use the set lifting of s defined in the obvious way: for $X \subseteq M$, $\mathsf{s}(X) = \{\mathsf{s}(x) \mid x \in X\}$. We also write $xy$ instead of $x \otimes y$.

**Lemma 3.** *The following holds for each multimonoid $M$: 1. $\mathsf{s}(x)\mathsf{s}(x) = \mathsf{s}(x)$; 2. $\mathsf{s}(\mathsf{s}(x)y) = \mathsf{s}(x)\mathsf{s}(y)$; 3. $\mathsf{s}(xy) \subseteq \mathsf{s}(x\mathsf{s}(y))$. If $M$ is local, then 4. $\mathsf{s}(x\mathsf{s}(y)) \subseteq \mathsf{s}(xy)$.*

Now we turn to the proof of Proposition 1.

*Proof.* (1.) If $h \in Di(Q^M)$, then $\mathsf{d}(h) = h$. Note that $\mathsf{d}(h)(x) = \bigvee_{x=\mathsf{s}(y)} h(y) = \bigvee_{x=\mathsf{s}(y) \& y \in I} h(y)$ (the last equality holds since $h \in Di(Q^M)$). However, $\mathsf{s}(y) = y$ by Lemma 2(1), and so $\bigvee_{x=\mathsf{s}(y) \& y \in I} h(y) = h(x)$.

(2.) $\mathsf{d}(f \cdot g) \le \mathsf{d}(f \cdot \mathsf{d}(g))$. We reason as follows:

$$\mathsf{d}(f \cdot \mathsf{d}(g))(x) = \bigvee_{x=\mathsf{s}(y)} \bigvee_{y \in z \otimes u} (f(z) \cdot \mathsf{d}(g)(u)) \quad = \bigvee_{x \in \mathsf{s}(z \otimes u)} \left( f(z) \cdot \bigvee_{u=\mathsf{s}(v)} g(v) \right)$$

$$= \bigvee_{x \in \mathsf{s}(z \otimes \mathsf{s}(v))} (f(z) \cdot g(v)) \overset{\text{Lemma 3(3)}}{\ge} \bigvee_{x \in \mathsf{s}(z \otimes v)} (f(z) \cdot g(v))$$

$$= \mathsf{d}(f \cdot g)(x)$$

We note that (5.) is established in a similar fashion using Lemma 3(3).

(3.) $\mathsf{d}(\mathsf{d}(f) \cdot g) = \mathsf{d}(f) \cdot \mathsf{d}(g)$. We reason as follows:

$$\mathsf{d}(\mathsf{d}(f) \cdot g)(x) = \bigvee_{x \in \mathsf{s}(y \otimes z)} (\mathsf{d}(f)(y) \cdot g(z)) \quad = \bigvee_{x \in \mathsf{s}(y \otimes z) \& y \in I} (\mathsf{d}(f)(y) \cdot g(z))$$

$$\overset{\text{Lemma 2(1)}}{=} \bigvee_{x \in \mathsf{s}(\mathsf{s}(x) \otimes z)} (\mathsf{d}(f)(x) \cdot g(z))$$

$$\overset{\text{Lemma 3(1, 2)}}{=} \left( \mathsf{d}(f)(x) \cdot \bigvee_{x=\mathsf{s}(z)} g(z) \right) \quad = \mathsf{d}(f)(x) \cdot \mathsf{d}(g)(x)$$

$$\overset{\text{Lemma 2(1–3)}}{=} (\mathsf{d}(f) \cdot \mathsf{d}(g))(x)$$

(4.) $\mathsf{d}\left( \bigvee_{i \in I} f_i \right) = \left( \bigvee_{i \in I} \mathsf{d}(f_i) \right)$. This is established as follows:

$$\mathsf{d}\left( \bigvee_{i \in I} f_i \right)(x) = \bigvee_{x=\mathsf{s}(y)} \left( \bigvee_{i \in I} f_i \right)(y) \quad = \bigvee_{x=\mathsf{s}(y)} \left( \bigvee_{i \in I} f_i(y) \right)$$

$$= \bigvee_{i \in I} \left( \bigvee_{x=\mathsf{s}(y)} f_i(y) \right) \quad = \bigvee_{i \in I} (\mathsf{d}(f_i)(x)) \quad = \left( \bigvee_{i \in I} \mathsf{d}(f_i) \right)(x)$$

$\square$

## A.3  Proof of Proposition 2

*Proof.* (1.) $\mathsf{d}(f) \le 1$ is not valid: it is sufficient to consider a *non-integral* $Q$ (1 is not the top element). (2.) $\mathsf{d}(f) \cdot f = f$ is not valid: consider the pair multimonoid $\{w\} \times \{w\}$ and the *non-idempotent* product quantale $\Pi$ where $f(w, w) = 0.5$. $\square$

## A.4     Proof of Proposition 3

*Proof.* We reason as follows:

$$\mathsf{d}(f \cdot g)(x) = \bigvee_{x=\mathsf{s}(y)} ((f \cdot g)(y)) \quad = \quad \bigvee_{x=\mathsf{s}(y)} \left( \bigvee_{y \in u \otimes v} (f(u) \cdot g(v)) \right)$$

$$= \bigvee_{x=\mathsf{s}(y)} \left( \bigvee_{y \in u \otimes v \& v \in I} (f(u) \cdot g(v)) \right) \quad = \quad \bigvee_{x=\mathsf{s}(y)} (f(y) \cdot g(\mathsf{t}(y)))$$

The first two equalities hold by definition. The third equality holds since $g$ is diagonal. The fourth equality is established as follows: $y \in y \otimes \mathsf{t}(y)$, and so the right hand side is less or equal than the left hand side; conversely, if $y \in u \otimes v \& v \in I$, then $y = u$ by Lemma 2(1) and $v = \mathsf{t}(y)$ since $\mathsf{t}(y)$ is the unique right identity of $y$ by definition. Hence, the left hand side is less or equal than the right hand side.                                  □

# References

1. Apt, K.R., de Boer, F.S., Olderog, E.R.: Verification of Sequential and Concurrent Programs, 3rd edn. Texts in Computer Science, Springer (2009)
2. Batz, K., Gallus, A., Kaminski, B.L., Katoen, J.P., Winkler, T.: Weighted programming: a programming paradigm for specifying mathematical models. Proc. ACM Program. Lang. **6**(OOPSLA1), 1–30 (2022). https://doi.org/10.1145/3527310
3. Desharnais, J., Möller, B., Struth, G.: Kleene algebra with domain. ACM Trans. Comput. Logic **7**(4), 798–833 (2006). https://doi.org/10.1145/1183278.1183285
4. Desharnais, J., Struth, G.: Internal axioms for domain semirings. Sci. Comput. Program. **76**(3), 181–203 (2011). https://doi.org/10.1016/j.scico.2010.05.007
5. Dijkstra, E.W.: Guarded commands, nondeterminacy and formal derivation of programs. Commun. ACM **18**(8), 453–457 (1975). https://doi.org/10.1145/360933. 360975
6. Fahrenberg, U., Johansen, C., Struth, G., Ziemiański, K.: Catoids and modal convolution algebras. Algebra Univ. **84**(2), 2023. https://doi.org/10.1007/s00012-023-00805-9
7. Fischer, M.J., Ladner, R.E.: Propositional dynamic logic of regular programs. J. Comput. Syst. Sci. **18**, 194–211 (1979). https://doi.org/10.1016/0022-0000(79)90046-1
8. Galatos, N., Jipsen, P., Kowalski, T., Ono, H.: Residuated Lattices: An Algebraic Glimpse at Substructural Logics. Elsevier (2007)
9. Gomes, L., Madeira, A., Barbosa, L.S.: On Kleene algebras for weighted computation. In: Cavalheiro, S., Fiadeiro, J. (eds.) Formal Methods: Foundations and Applications, pp. 271–286. Springer International Publishing, Cham (2017). https://doi.org/10.1007/978-3-319-70848-5_17
10. Gomes, L., Madeira, A., Barbosa, L.S.: Generalising KAT to verify weighted computations. Sci. Ann. Comput. Sci. **29**(2), 141–184 (2019). https://doi.org/10.7561/SACS.2019.2.141
11. Groenendijk, J., Stokhof, M.: Dynamic predicate logic. Linguist. Philos. **14**(1), 39–100 (1991). https://doi.org/10.1007/BF00628304

12. Hoare, C.A.R.: An axiomatic basis for computer programming. Commun. ACM **12**, 576–580 (1969). https://doi.org/10.1145/363235.363259
13. Kozen, D.: On Kleene algebras and closed semirings. In: Rovan, B. (ed.) International Symposium on Mathematical Foundations of Computer Science, pp. 26–47. Springer (1990). https://doi.org/10.1007/BFb0029594
14. Kozen, D.: A completeness theorem for Kleene algebras and the algebra of regular events. Inf. Comput. **110**(2), 366–390 (1994). https://doi.org/10.1006/inco.1994.1037
15. Kozen, D.: Kleene algebra with tests. ACM Trans. Program. Lang. Syst. **19**(3), 427–443 (1997). https://doi.org/10.1145/256167.256195
16. Kozen, D., Smith, F.: Kleene algebra with tests: completeness and decidability. In: Dalen, D., Bezem, M. (eds.) Computer Science Logic, pp. 244–259. Springer Berlin Heidelberg, Berlin, Heidelberg (1997). https://doi.org/10.1007/3-540-63172-0_43
17. Kudryavtseva, G., Mazorchuk, V.: On multisemigroups. Portugaliae. Mathematica **72**(1), 47–80 (2015). https://doi.org/10.4171/pm/1956
18. Paseka, J., Rosický, J.: Quantales. In: Coecke, B., Moore, D., Wilce, A. (eds.) Current Research in Operational Quantum Logic: Algebras, Categories, Languages, pp. 245–262. Springer Netherlands, Dordrecht (2000). https://doi.org/10.1007/978-94-017-1201-9_10
19. Restall, G.: An Introduction to Substrucutral Logics. Routledge, London (2000)
20. Rosenthal, K.I.: Quantales and Their Applications. Longman Scientific & Technical, (1990)
21. Routley, R., Plumwood, V., Meyer, R.K., Brady, R.T.: Relevant Logics and Their Rivals, vol. 1. Ridgeview (1982)
22. Różowski, W., Silva, A.: A Completeness Theorem for Probabilistic Regular Expressions. Technical report (2023). https://doi.org/10.48550/ARXIV.2310.08779
23. Różowski, W., Kappé, T., Kozen, D., Schmid, T., Silva, A.: Probabilistic Guarded KAT modulo bisimilarity: Completeness and complexity. In: Etessami, K., Feige, U. and Puppis, G. (eds.) Proc. 50th International Colloquium on Automata, Languages, and Programming (ICALP 2023), pp. 136:1–136:20. No. 261 in Leibniz International Proceedings in Informatics (LIPIcs), Schloss Dagstuhl - Leibniz-Zentrum für Informatik, Dagstuhl, Germany (2023). https://doi.org/10.4230/LIPIcs.ICALP.2023.136
24. Sedlár, I.: Kleene algebra with tests for weighted programs. In: Proceedings of the IEEE 53rd International Symposium on Multiple-Valued Logic (ISMVL 2023), pp. 111–116 (2023). https://doi.org/10.1109/ISMVL57333.2023.00031
25. Sedlár, I.: On the complexity of Kleene algebra with domain. In: Gl überück, R., Santocanale, L., Winter, M. (eds.) Relational and Algebraic Methods in Computer Science (RAMiCS 2023). LNCS, vol. 13896, pp. 208–223. Springer, Cham (2023). https://doi.org/10.1007/978-3-031-28083-2_13
26. Smolka, S., Foster, N., Hsu, J., Kappé, T., Kozen, D., Silva, A.: Guarded Kleene algebra with tests: Verification of uninterpreted programs in nearly linear time. In: Proceedings 47th ACM SIGPLAN Symp. Principles of Programming Languages (POPL'20), pp. 61:1–28. ACM, New Orleans (2020). https://doi.org/10.1145/3371129

# Automated Quantum Program Verification in Dynamic Quantum Logic

Tsubasa Takagi[✉][iD], Canh Minh Do[iD], and Kazuhiro Ogata[iD]

Japan Advanced Institute of Science and Technology, Nomi, Japan
{tsubasa,canhdo,ogata}@jaist.ac.jp

**Abstract.** Dynamic Quantum Logic (DQL) has been used as a logical framework for manually proving the correctness of quantum programs. This paper presents an automated approach to quantum program verification at the cost of simplifying DQL to Basic Dynamic Quantum Logic (BDQL). We first formalize quantum states, quantum gates, and projections in bra-ket notation and use a set of laws from quantum mechanics and matrix operations to reason on quantum computation. We then formalize the semantics of BQDL and specify the behavior and desired properties of quantum programs in the scope of BDQL. Formal verification of whether a quantum program satisfies desired properties is conducted automatically through an equational simplification process. We use Maude, a rewriting logic-based specification/programming language, to implement our approach. To demonstrate the effectiveness of our automated approach, we successfully verified the correctness of five quantum protocols: Superdense Coding, Quantum Teleportation, Quantum Secret Sharing, Entanglement Swapping, and Quantum Gate Teleportation, using our support tool.

**Keywords:** Dynamic Quantum Logic · Quantum Programs · Quantum Protocols · Maude

## 1 Introduction

Quantum computing has the potential to transform various computing applications, such as cryptography [24], deep learning [8], optimization [12], and solving linear systems [17], by offering the ability to solve problems that are currently infeasible for classical computing, such as Shore's fast algorithm for integer factoring and Grover's fast algorithm for finding a datum in an unsorted database. However, quantum computing is counter-intuitive and distinct from classical computing, which makes it challenging to design and implement quantum protocols, algorithms, and programs accurately. Therefore, it is crucial to ensure their correctness through verification. While existing formal verification techniques can be used to verify that classical systems enjoy some desired properties, they cannot be directly applied to quantum systems due to the distinct principles used in quantum computing [27]. Therefore, new formal verification techniques are necessary for quantum systems.

N. Gierasimczuk and F. R. Velázquez-Quesada (Eds.): DaLí 2023, LNCS 14401, pp. 68–84, 2024.
https://doi.org/10.1007/978-3-031-51777-8_5

An extension of Quantum Logic [9] called Dynamic Quantum Logic (DQL) can be utilized to describe specifications of quantum programs. So far, various quantum protocols, such as Superdense Coding [6], Quantum Teleportation [5], Quantum Secret Sharing [19], Entanglement Swapping [28], and Quantum Gate Teleportation [14] have been verified using a DQL called the Logic of Quantum Programs (LQP) [1,2] (see [4] for a comprehensive review of DQL). However, these protocols were only verified manually by giving their correctness proofs, and it has not been known how to automate this process.

This paper presents an automated approach to quantum program verification in Basic Dynamic Quantum Logic (BDQL). BDQL is a simplified version of DQL, reflecting its essential features from an implementation perspective. We first formalize quantum states, quantum gates, and projections in bra-ket notation and use a set of laws from quantum mechanics and matrix operations to reason on quantum computation. This formalization is adopted from symbolic reasoning in [11]. The advantage of this symbolic reasoning is that we use bra-ket notation instead of explicitly complex vectors and matrices as is proposed in [22], which makes our representations more compact. Moreover, we can deal not only with concrete values but also with symbolic values for complex numbers reasoning. We then formalize the semantics of BDQL in order to describe the behavior and desired properties of quantum programs. We use Maude [10], a high-performance specification/programming language based on rewriting logic [20], to implement our approach. The symbolic reasoning on quantum computation and the semantics of BDQL are formalized by means of equations in Maude. Therefore, formal verification of quantum programs in BDQL is conducted automatically through a simplification process with respect to the equations in Maude.

Using our support tool, we successfully verify the correctness of five quantum protocols: Superdense Coding, Quantum Teleportation, Quantum Secret Sharing, Entanglement Swapping, and Quantum Gate Teleportation. This demonstrates the effectiveness of our automated approach to verify quantum programs in BDQL with symbolic reasoning adopted from [11] in practice. The support tool and case studies are available at https://github.com/canhminhdo/DQL.

## 2   Basic Dynamic Quantum Logic

In this section, we formulate Basic Dynamic Quantum Logic (BDQL). It is possible to describe and verify at least the five specific protocols in Sect. 3. Because the five protocols are utilized for more complex protocols, BDQL is a sufficiently expressive logic as a starting point. Further extensions of our BDQL will be required to verify other protocols in the future.

Let $L_0$ be a set of atomic formulas and $\Pi_0$ be a set of atomic programs. The set $L$ of all formulas in BDQL and the set $\Pi$ of all star-free regular programs are generated by simultaneous induction as follows:

$$L \ni A ::= p \mid \neg A \mid A \wedge A \mid [a]A,$$
$$\Pi \ni a ::= \mathbf{skip} \mid \mathbf{abort} \mid \pi \mid a \mathbin{;} a \mid a \cup a \mid A?,$$

where $p \in L_0$ and $\pi \in \Pi_0$. The symbols **skip** and **abort** are called constant programs. The operators ;, $\cup$, and ? are called sequential composition, non-deterministic choice, and test, respectively.

The syntax of BDQL is exactly the same as that of Propositional Dynamic Logic (PDL) [16] without the Kleene star operator *. It is not strange that two different logics have the same syntax. For example, Classical Logic and Intuitionistic Logic have the same syntax but are distinguished by their semantics (or their sets of provable formulas). Similarly, the semantics of BDQL and that of the star-free fragment of PDL are different. This paper considers only star-free regular programs, leaving the addition of * to the logic in a future paper.

We define the semantics of BDQL using frames and models as usual. This kind of semantics is called Kripke (or relational) semantics.

– A quantum dynamic frame is a pair $F = (\mathcal{H}, v)$ that consists of a Hilbert space $\mathcal{H}$ and a function $v$ from $\Pi_0$ to the set $\mathcal{U}(\mathcal{H})$ of all unitary operators on $\mathcal{H}$. The function $v$ is called an interpretation for atomic programs.
– A quantum dynamic model is a triple $M = (\mathcal{H}, v, V)$ that consists of a quantum dynamic frame $(\mathcal{H}, v)$ and a function $V$ from $L_0$ to the set $\mathcal{C}(\mathcal{H})$ of all closed subspaces of $\mathcal{H}$. The function $V$ is called an interpretation for atomic formulas.

The definition of quantum dynamic models states that atomic formulas are interpreted as a closed subspace of a Hilbert space. This interpretation is known as the algebraic semantics for Quantum Logic [9]. The set $\mathcal{C}(\mathcal{H})$ is called a Hilbert lattice [23] because it forms a lattice with meet $X \cap Y$ and join $X \sqcup Y = (X^{\perp} \cap Y^{\perp})^{\perp}$ for any $X, Y \in \mathcal{C}(\mathcal{H})$, where $\perp$ denotes the orthogonal complement. Note that $X \sqcup Y \supseteq X \cup Y$ and $X \cup Y \notin \mathcal{C}(\mathcal{H})$ in general.

*Remark 1.* Usually, Kripke frames are defined as a pair (tuple) that consists of a non-empty set $S$ and relation(s) $R$ on $S$. On the other hand, the quantum dynamic frames defined above have no relation(s). However, the relations can be recovered immediately using $v$. That is, the family $\{R_{\pi} : \pi \in \Pi_0\}$ of relations on $\mathcal{H}$ is constructed by

$$R_{\pi} = \{(s, t) : (v(\pi))(s) = t\}$$

for each $\pi \in \Pi_0$. For this reason, we use the word "frame" for quantum dynamic frames.

The interpretation $v$ is defined for atomic programs, and $V$ is defined for atomic formulas. These interpretations are extended to that for star-free regular programs and formulas, respectively. For any quantum dynamic model $M$, the function $[\![\ ]\!]^M : L \to \mathcal{C}(\mathcal{H})$ and family $\{R_a^M : a \in \Pi\}$ of relations on $\mathcal{H}$ are defined by simultaneous induction as follows:

1. $[\![p]\!]^M = V(p)$;
2. $[\![\neg A]\!]^M$ is the orthogonal complement of $[\![A]\!]^M$;
3. $[\![A \wedge B]\!]^M = [\![A]\!]^M \cap [\![B]\!]^M$;

4. $[\![a]A]\!]^M = \{s \in \mathcal{H} : (s,t) \in R_a^M \text{ implies } t \in [\![A]\!]^M \text{ for any } t \in \mathcal{H}\}$;
5. $R_{\mathbf{skip}}^M = \{(s,t) : s = t\}$;
6. $R_{\mathbf{abort}}^M = \emptyset$;
7. $R_\pi^M = \{(s,t) : (v(\pi))(s) = t\}$;
8. $R_{a;b}^M = \{(s,t) : (s,u) \in R_a^M \text{ and } (u,t) \in R_b^M \text{ for some } u \in \mathcal{H}\}$;
9. $R_{a \cup b}^M = R_a^M \cup R_b^M$;
10. $R_{A?}^M = \{(s,t) : P_{[\![A]\!]^M}(s) = t\}$, where $P_{[\![A]\!]^M}$ stands for the projection onto $[\![A]\!]^M$.

**Theorem 1.** $[\![\ ]\!]^M$ *is well-defined. That is,* $[\![A]\!]^M \in \mathcal{C}(\mathcal{H})$ *for each* $A \in L$.

*Proof.* See Appendix.

The function $[\![\ ]\!]^M$ and family $\{R_a^M : a \in \Pi\}$ are uniquely determined if $M$ is given. Recall that $v(\pi)$ is a function. On the other hand, $R_a^M$ is a relation and may not be a function due to $\cup$.

Now we can understand the meaning of each program: **skip** does nothing, **abort** forces to halt without executing subsequent programs, ; is the composition operator, $\cup$ is the non-deterministic choice operator, and ? is the quantum test operator and is used to represent a result of projective measurement (see Sect. 3.1).

Henceforth, we write $(M,s) \models A$ for the condition $s \in [\![A]\!]^M$ as usual. That is, $(M,s) \models A$ if and only if $P_{[\![A]\!]^M}(s) = s$. A formula $A$ is said to be satisfiable (resp. valid) if $(M,s) \models A$ for some (resp. any) $M$ and $s \in \mathcal{H}$.

*Remark 2.* In most modal logics, a contradiction $A \wedge \neg A$ is not satisfiable. In other words, not $(M,s) \models A \wedge \neg A$ for any $s$. On the other hand, $A \wedge \neg A$ is satisfiable in BDQL because $(M,\mathbf{0}) \models A \wedge \neg A$, where $\mathbf{0}$ stands for the origin (zero vector) of $\mathcal{H}$. LQP [2] chooses different semantics from that in this paper to avoid this. That is, $\mathbf{0}$ (or the corresponding subspace $\{\mathbf{0}\}$) is not a state in the semantics of LQP. Unlike LQP, we allow $\mathbf{0}$ to be a state; otherwise, our definition is ill-defined (Theorem 1 does not hold).

The following theorem gives the theoretical background for rewriting the statement of the form $(M,s) \models A$ in implementation explained in Sect. 4.

**Theorem 2.** *The following holds for any $M$ and $s \in \mathcal{H}$.*

1. $(M,s) \models A \wedge B$, *if and only if* $(M,s) \models A$ *and* $(M,s) \models B$.
2. $(M,s) \models [\mathbf{skip}]A$ *if and only if* $(M,s) \models A$.
3. $(M,s) \models [\mathbf{abort}]A$.
4. $(M,s) \models [\pi]A$ *if and only if* $(M,(v(\pi))(s)) \models A$.
5. $(M,s) \models [a ; b]A$ *if and only if* $(M,s) \models [a][b]A$.
6. $(M,s) \models [a \cup b]A$ *if and only if* $(M,s) \models [a]A \wedge [b]A$.
7. $(M,s) \models [A?]B$ *if and only if* $(M, P_{[\![A]\!]^M}(s)) \models B$.

*Proof.* Straightforward.

# 3  Application to Quantum Program Verification

This section describes the behavior and desired properties of some specific quantum programs in the language of BDQL. These properties can be verified automatically using our support tool as shown in Sect. 4 and 5.

## 3.1  Basic Notions

In the beginning, we briefly review quantum computation and fix our notation. We assume the readers have basic knowledge of linear algebra.

Generally speaking, quantum systems are formulated as complex Hilbert spaces. However, for quantum computation, it is enough to consider specific Hilbert spaces called qubit systems. An $n$-qubit system is the complex $2^n$-space $\mathbb{C}^{2^n}$, where $\mathbb{C}$ stands for the complex plane. Pure states in the $n$-qubit system $\mathbb{C}^{2^n}$ are unit vectors in $\mathbb{C}^{2^n}$. The orthogonal basis called computational basis in the one-qubit system $\mathbb{C}^2$ is a set $\{|0\rangle, |1\rangle\}$ that consists of the column vectors $|0\rangle = (1,0)^T$ and $|1\rangle = (0,1)^T$, where $T$ denotes the transpose operator. The linear combinations $|+\rangle = (|0\rangle + |1\rangle)/\sqrt{2}$ and $|-\rangle = (|0\rangle - |1\rangle)/\sqrt{2}$ of $|0\rangle$ and $|1\rangle$ are also pure states. In general, $|\psi\rangle = c_0 |0\rangle + c_1 |1\rangle$ represents a pure state in the one-qubit system $\mathbb{C}^2$ provided that $|c_0|^2 + |c_1|^2 = 1$. In the two-qubit system $\mathbb{C}^4$, there are pure states that cannot be represented in the form $|\psi_1\rangle \otimes |\psi_2\rangle$ and are called entangled states, where $\otimes$ denotes the tensor product (more precisely, the Kronecker product). For example, the EPR state (Einstein-Podolsky-Rosen state) $|\text{EPR}\rangle = (|00\rangle + |11\rangle)/\sqrt{2}$ is an entangled state, where $|00\rangle = |0\rangle \otimes |0\rangle$ and $|11\rangle = |1\rangle \otimes |1\rangle$.

The above notation of vectors is called bra-ket notation (also called Dirac notation). $|\psi\rangle$ is called a ket vector. A bra vector $\langle\psi|$ is defined as a row vector whose elements are complex conjugates of the elements of the corresponding ket vector $|\psi\rangle$. Observe that the matrix multiplication $|\psi\rangle\langle\psi|$ is the projection onto the subspace spanned by $|\psi\rangle$.

Quantum computation is represented by unitary operators (also called quantum gates). There are various quantum gates. For example, the Hadamard gate $H$ and Pauli gates $X$, $Y$, and $Z$ are typical quantum gates on the one-qubit system $\mathbb{C}^2$ and are defined as follows:

$$H = \frac{1}{\sqrt{2}}\begin{pmatrix} 1 & 1 \\ 1 & -1 \end{pmatrix}, \quad X = \begin{pmatrix} 0 & 1 \\ 1 & 0 \end{pmatrix}, \quad Y = \begin{pmatrix} 0 & -i \\ i & 0 \end{pmatrix}, \quad Z = \begin{pmatrix} 1 & 0 \\ 0 & -1 \end{pmatrix}.$$

Two typical quantum gates on the two-qubit system $\mathbb{C}^4$ are the controlled-X gate (also called the controlled NOT gate) CX and the swap gate SWAP are defined by

$$\text{CX} = |0\rangle\langle0| \otimes I + |1\rangle\langle1| \otimes X,$$
$$\text{SWAP} = \text{CX}(I \otimes |0\rangle\langle0| + X \otimes |1\rangle\langle1|)\text{CX},$$

where $I$ denotes the identity matrix of size $2 \times 2$. Measurement is a completely different process from applying quantum gates. Here, we roughly explain specific projective measurements. For the general definition of projective measurement, see [21]. Observe that $P_0 = |0\rangle\langle 0|$ and $P_1 = |1\rangle\langle 1|$ are projections, respectively. After executing the measurement $\{P_0, P_1\}$, a current state $|\psi\rangle = c_0 |0\rangle + c_1 |1\rangle$ is transitioned into $P_0 |\psi\rangle / |c_0| = c_0 |0\rangle / |c_0|$ with probability $|c_0|^2$ and into $P_1 |\psi\rangle / |c_1| = c_1 |1\rangle / |c_1|$ with probability $|c_1|^2$. There is no other possibility because $|c_0|^2 + |c_1|^2 = 1$.

## 3.2 Standard Interpretation

To describe the quantum programs discussed in this paper, we fix

$$\Pi_0 = \{\text{H}(i), \text{X}(i), \text{Y}(i), \text{Z}(i), \text{CX}(i, j), \text{SWAP}(i, j) : i, j \in \mathbb{N}, i \neq j\},$$
$$L_0 = \{p(i, |\psi\rangle), p(i, i+1, |\Psi\rangle) : i \in \mathbb{N}, |\psi\rangle \in \mathbb{C}^2, |\Psi\rangle \in \mathbb{C}^4\},$$

where $\mathbb{N}$ stands for the set of all natural numbers (including 0). Because now atomic programs and atomic formulas are restricted, we only need to consider specific interpretations called the standard interpretations $\bar{v}$ and $\overline{V}$ instead of $v$ and $V$, respectively. The standard interpretations are defined below.

A function $\bar{v} : \Pi_0 \to \mathcal{U}(\mathbb{C}^{2^n})$ is called the standard interpretation on $\mathbb{C}^{2^n}$ for atomic programs if

$$\bar{v}(\text{H}(i)) = I^{\otimes i} \otimes H \otimes I^{\otimes n-i-1}, \quad \bar{v}(\text{X}(i)) = I^{\otimes i} \otimes X \otimes I^{\otimes n-i-1},$$
$$\bar{v}(\text{Y}(i)) = I^{\otimes i} \otimes Y \otimes I^{\otimes n-i-1}, \quad \bar{v}(\text{Z}(i)) = I^{\otimes i} \otimes Z \otimes I^{\otimes n-i-1},$$

$$\bar{v}(\text{CX}(i, j)) = I^{\otimes i} \otimes |0\rangle\langle 0| \otimes I^{\otimes n-i-1}$$
$$+ (I^{\otimes i} \otimes |1\rangle\langle 1| \otimes I^{\otimes n-i-1})(I^{\otimes j} \otimes X \otimes I^{\otimes n-j-1}),$$
$$\bar{v}(\text{SWAP}(i, j)) = \bar{v}(\text{CX}(i, j) \; ; \; \text{CX}(j, i) \; ; \; \text{CX}(i, j)),$$

where

$$I^{\otimes i} = \overbrace{I \otimes \cdots \otimes I}^{i}.$$

That is, under the standard interpretation, $\text{H}(i)$, $\text{X}(i)$, $\text{Y}(i)$, $\text{Z}(i)$ execute the corresponding quantum gate on the $i$-th qubit, $\text{CX}(i, j)$ executes the Pauli gate $X$ on the target qubit ($j$-th qubit) depending on the state of the control qubit ($i$-th qubit), and $\text{SWAP}(i, j)$ swaps the $i$-th and $j$-th qubits.

A function $\overline{V} : L_0 \to \mathcal{C}(\mathbb{C}^{2^n})$ is called the standard interpretation on $\mathbb{C}^{2^n}$ for atomic formulas if

$$\overline{V}(p(i, |\psi\rangle)) = \mathbb{C}^{2^i} \otimes \text{span}\{|\psi\rangle\} \otimes \mathbb{C}^{2^{n-i-1}},$$
$$\overline{V}(p(i, i+1, |\Psi\rangle)) = \mathbb{C}^{2^i} \otimes \text{span}\{|\Psi\rangle\} \otimes \mathbb{C}^{2^{n-i-2}},$$

where $\text{span}\{|\psi\rangle\}$ (resp. $\text{span}\{|\Psi\rangle\}$) stands for the subspace spanned by $\{|\psi\rangle\}$ (resp. $\{|\Psi\rangle\}$).

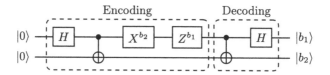

**Fig. 1.** Superdense Coding

In what follows, we write $\overline{M}_n$ for $(\mathbb{C}^{2^n}, \bar{v}, \overline{V})$, where the index $n$ represents the number of qubits. In addition, we use the following abbreviation to conventionally describe quantum programs in BDQL:

$$\textbf{if } A \textbf{ then } a \textbf{ else } b \textbf{ fi} = (A? \; ; a) \cup (\neg A? \; ; b).$$

This program means the selection depends on the outcomes of projective measurement. That is, execute $a$ or $b$ depending on the result of the measurement $\{P_{\lceil A \rceil M}, P_{\lceil \neg A \rceil M}\}$. Because projective measurement occurs only in quantum computation, the behavior of this selection command is different from the usual (classical) **if then else fi** program.

### 3.3   Case Studies

#### Superdense Coding

Superdense Coding [6] allows us to transmit two classical bits using an entangled state. It consists of encoding and decoding the information. The encoding process of information 00, 01, 10, or 11 is described as follows:

$$\textbf{encode}_{00} = \texttt{H}(0) \; ; \texttt{CX}(0,1), \quad \textbf{encode}_{01} = \texttt{H}(0) \; ; \texttt{CX}(0,1) \; ; \texttt{X}(0),$$

$$\textbf{encode}_{10} = \texttt{H}(0) \; ; \texttt{CX}(0,1) \; ; \texttt{Z}(0), \quad \textbf{encode}_{11} = \texttt{H}(0) \; ; \texttt{CX}(0,1) \; ; \texttt{X}(0) \; ; \texttt{Z}(0).$$

The decoding process is described as $\textbf{decode} = \texttt{CX}(0,1) \; ; \texttt{H}(0)$.

The desired property for Superdense Coding is that "the encoded information is correctly decoded." In BDQL, this property is expressed as follows:

$$(\overline{M}_2, |0\rangle \otimes |0\rangle) \models \bigwedge_{i,j \in \{0,1\}} [\textbf{encode}_{ij} \; ; \textbf{decode}](p(0,|i\rangle) \wedge p(1,|j\rangle)).$$

#### Quantum Teleportation

Quantum Teleportation [5] is a protocol for teleporting an arbitrary pure state by sending two bits of classical information. The program of Quantum Teleportation is described as follows:

$$\begin{aligned}
\textbf{teleport} = \; &\texttt{H}(1) \; ; \texttt{CX}(1,2) \; ; \texttt{CX}(0,1) \; ; \texttt{H}(0) \\
&; \textbf{if } p(1,|0\rangle) \textbf{ then skip else } \texttt{X}(2) \textbf{ fi} \\
&; \textbf{if } p(0,|0\rangle) \textbf{ then skip else } \texttt{Z}(2) \textbf{ fi}.
\end{aligned}$$

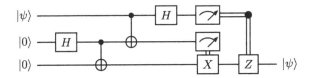

**Fig. 2.** Quantum Teleportation

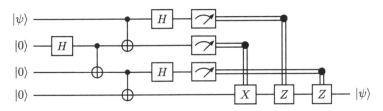

**Fig. 3.** Quantum Secret Sharing

The desired property of Quantum Teleportation is that "a pure state $|\psi\rangle$ is correctly teleported." In BDQL, this property is expressed as follows:

$$(\overline{M}_3, |\psi\rangle \otimes |0\rangle \otimes |0\rangle) \models [\textbf{teleport}]p(2, |\psi\rangle).$$

## Quantum Secret Sharing

Quantum Secret Sharing [19] is a protocol for teleporting a pure state from a sender (Alice) to a receiver (Bob) with the help of a third party (Charlie). By this protocol, a secret pure state is shared between Alice and Bob, provided that Charlie permits it. The program of Quantum Secret Sharing is described as follows:

$$\textbf{share} = \texttt{H}(1) \; ; \; \texttt{CX}(1,2) \; ; \; \texttt{CX}(0,1) \; ; \; \texttt{H}(0) \; ; \; \texttt{CX}(2,3) \; ; \; \texttt{H}(2)$$
$$; \textbf{ if } p(1, |0\rangle) \textbf{ then skip else } \texttt{X}(3)) \textbf{ fi}$$
$$; \textbf{ if } p(0, |0\rangle) \textbf{ then skip else } \texttt{Z}(3) \textbf{ fi}$$
$$; \textbf{ if } p(2, |0\rangle) \textbf{ then skip else } \texttt{Z}(3) \textbf{ fi}.$$

The desired property of secret sharing is similar to that of Quantum Teleportation. In BDQL, this property is expressed as follows:

$$(\overline{M}_4, |\psi\rangle \otimes |0\rangle \otimes |0\rangle \otimes |0\rangle) \models [\textbf{share}]p(3, |\psi\rangle)).$$

## Entanglement Swapping

Entanglement Swapping [28] is a protocol for creating a new entangled state. Suppose that Alice and Bob share two entangled qubits, and Bob and Charlie also share two different entangled qubits. After executing Entanglement Swapping, Alice's qubit and Charlie's qubit become entangled. The program of Entanglement Swapping is described as follows:

$$\textbf{entangle} = \texttt{H}(0) ; \texttt{CX}(0,1) ; \texttt{H}(2) ; \texttt{CX}(2,3) ; \texttt{CX}(1,2) ; \texttt{H}(1)$$
$$; \textbf{if } p(2,|0\rangle) \textbf{ then skip else } \texttt{X}(3) \textbf{ fi}$$
$$; \textbf{if } p(1,|0\rangle) \textbf{ then skip else } \texttt{Z}(3) \textbf{ fi}$$
$$; \texttt{SWAP}(1,3).$$

The last $\texttt{SWAP}(1,3)$ is executed to adjoin the remaining qubits.

The desired property of Entanglement Swapping is that "an entangled state (in this case, $|\text{EPR}\rangle$) is created." In BDQL, this property is expressed as follows:

$$(\overline{M}_4, |0\rangle \otimes |0\rangle \otimes |0\rangle \otimes |0\rangle) \models [\textbf{entangle}]p(0,1,|\text{EPR}\rangle).$$

Note that $\texttt{SWAP}(1,3)$ is needed because $p(i, i+1, |\Psi\rangle)$ is only defined for the consecutive numbers $i$ and $i+1$. That is, the expression $p(0, 3, |\text{EPR}\rangle)$ is not defined.

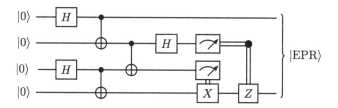

**Fig. 4.** Entanglement Swapping

## Quantum Gate Teleportation

Quantum Gate Teleportation [14] is a protocol for teleporting a quantum gate. The program of quantum gate teleportation is described as follows:

$$\textbf{gteleport} = \texttt{H}(1) ; \texttt{CX}(1,2) ; \texttt{H}(3) ; \texttt{CX}(3,4) ; \texttt{CX}(3,2) ; \texttt{CX}(0,1) ; \texttt{H}(0) ; \texttt{CX}(4,5) ; \texttt{H}(4)$$
$$; \textbf{if } p(0,|0\rangle) \textbf{ then skip else } \texttt{Z}(2) ; \texttt{Z}(3) \textbf{ fi}$$
$$; \textbf{if } p(1,|0\rangle) \textbf{ then skip else } \texttt{X}(2) \textbf{ fi}$$
$$; \textbf{if } p(5,|0\rangle) \textbf{ then skip else } \texttt{X}(2) ; \texttt{X}(3) \textbf{ fi}$$
$$; \textbf{if } p(4,|0\rangle) \textbf{ then skip else } \texttt{Z}(3) \textbf{ fi}.$$

The desired property of Quantum Gate Teleportation is that "a quantum gate (in this case, CX) is correctly teleported." In BDQL, this property is expressed as follows:

$$(\overline{M}_6, |\psi\rangle \otimes |0\rangle \otimes |0\rangle \otimes |0\rangle \otimes |0\rangle \otimes |\psi'\rangle) \models [\textbf{gteleport}]p(3,4,\text{CX}(|\psi'\rangle \otimes |\psi\rangle)).$$

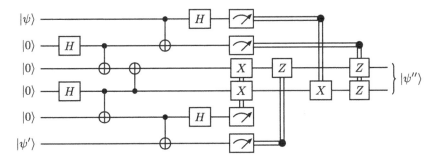

**Fig. 5.** Quantum Gate Teleportation ($|\psi''\rangle = \text{CX}(|\psi'\rangle \otimes |\psi\rangle)$))

## 4    Implementation of Basic Dynamic Quantum Logic

This section describes the implementation of BDQL in Maude [10], a specification/programming language based on rewriting logic [20]. Hence, the notations used in this section follow the Maude syntax.

### 4.1    Syntax of Basic Dynamic Quantum Logic

We formalize pure states in a program as a collection of qubits associated with indices that start from 0 to $N-1$, where $N$ is the total number of qubits. Hence, we can flexibly refer to a specific part of a quantum state using indices. We adopt this formalization to reason on quantum computation from [11].

We define two sorts AtomicProg and Prog for atomic programs $\Pi_0$ and star-free regular programs $\Pi$ in BDQL, respectively, where AtomicProg is a subsort of Prog. We also define several operators for atomic programs as follows:

```
sorts AtomicProg Prog .
subsort AtomicProg < Prog .
ops I(_) H(_) X(_) Y(_) Z(_) : Nat -> AtomicProg [ctor] .
op CX(_,_) : Nat Nat -> AtomicProg [ctor] .
```

where I(_), H(_), X(_), Y(_), Z(_) operators take a natural number as input, denoting the index of a qubit of a pure state on which the quantum gates $I$, $H$, $X$, $Y$, $Z$, will be applied, respectively; and CX(_,_) operator takes two natural numbers as inputs, denoting the indices of two qubits of a pure state on which CX will be applied. These operators serve as the constructors of atomic programs with the ctor attribute.

We use several operators to define star-free regular programs in BDQL as follows:

```
ops abort skip : -> Prog [ctor] .
op _;_ : Prog Prog -> Prog [ctor assoc id: skip prec 25] .
op _U_ : Prog Prog -> Prog [ctor] .
op _? : Formula -> Prog [ctor prec 24] .
```

where all operators follow the definition of star-free regular programs in BDQL shown in Sect. 2; besides that, the skip operator also denotes an empty program; ctor, assoc, id:_, and prec_ are operator attributes for a constructor, associativity, an identity element, and operator precedence, respectively.

We define two sorts AtomicFormula and Formula for atomic formulas $L_0$ and general formulas $L$ in BDQL, respectively, where AtomicFormula is a subsort of Formula. We also define several operators for constructing formulas in DQL as follows:

```
sorts AtomicFormula Formula .
subsort AtomicFormula < Formula .
op P(_,_) : Nat Matrix -> AtomicFormula [ctor] .
op P(_,_,_) : Nat Nat Matrix -> AtomicFormula [ctor] .
op neg_ : Formula -> Formula .
op _/\_ : Formula Formula -> Formula [ctor comm assoc] .
op [_]_ : Prog Formula -> Formula [ctor] .
```

where Matrix is a family sort of quantum states, quantum gates, and projections because they can be expressed in terms of matrices. The P(_,_) and P(_,_,_) operators are atomic formulas representing projections of the forms $p(i, |\psi\rangle)$ and $p(i, j, |\Psi\rangle)$, respectively (see Sect. 3.2). The other operators follow the definition of formulas in BDQL as shown in Sect. 2.

The if-then-else-fi command corresponding to the program **if then else fi** is implemented as follows:

```
op if_then_else_fi : Formula Prog Prog -> Prog .
eq if F1:Formula then P1:Prog else P2:Prog fi
= (F1:Formula ? ; P1:Prog) U ((neg F2:Formula) ? ; P2:Prog) .
```

## 4.2  Semantics of Basic Dynamic Quantum Logic

The semantics of $(M, s) \models A$ in BDQL is represented by the term s |= A of sort Judgment with the following operator.

```
sort Judgment .
op _|=_ : QState Formula -> Judgment .
```

where sort QState represents quantum states (more precisely, pure states).

The satisfiability of $A \wedge B$ in BDQL is determined by that of $A$ and $B$ (Theorem 2), each of which is represented by a judgment. Hence, we need a sort to represent a set of judgments as follows:

```
sort JudgmentSet .    subsort Judgment < JudgmentSet .
op emptyJS : -> JudgmentSet [ctor] .
op _/\_ : JudgmentSet JudgmentSet -> JudgmentSet [ctor assoc
    comm id: emptyJS] .
```

where emptyJS is the empty set of judgments, and the operator _/_ serves as the constructor of the set (ctor), is associative (assoc), commutative (comm), and has the empty set as an identity element (id: emptyJS).

Now, we implement the semantics of BDQL using equations that simplify a judgment s |= A into the set of judgments as follows:

```
vars PROG PROG' : Prog .           vars Q Q' : QState .
vars N N1 N2 : Nat .               var M : Matrix .
vars Phi Psi : Formula .
ceq Q |= P(N, M) = emptyJS if (Q).P(N, M) == Q .
ceq Q |= P(N1, N2, M) = emptyJS if (Q).P(N1, N2, M) == Q .
eq neg P(N:Nat, |0>) = P(N:Nat, |1>) .
eq neg P(N:Nat, |1>) = P(N:Nat, |0>) .
eq Q |= Phi /\ Psi = (Q |= Phi) /\ (Q |= Psi) .
eq Q |= [skip] Phi = Q |= Phi .
eq Q |= [abort] Phi = emptyJS .
eq Q |= [I(N)] Phi = Q |= Phi .
ceq Q |= [H(N)] Phi = Q' |= Phi if Q' := (Q).H(N) .
ceq Q |= [X(N)] Phi = Q' |= Phi if Q' := (Q).X(N) .
ceq Q |= [Y(N)] Phi = Q' |= Phi if Q' := (Q).Y(N) .
ceq Q |= [Z(N)] Phi = Q' |= Phi if Q' := (Q).Z(N) .
ceq Q |= [CX(N1,N2)] Phi = Q' |= Phi if Q' := (Q).CX(N1,N2) .
ceq Q |= [PROG' ; PROG] Phi = Q |= [PROG']([PROG] Phi)
if PROG' =/= nil /\ PROG =/= nil .
eq Q |= [PROG' U PROG] Phi
= (Q |= [PROG']) Phi) /\ (Q |= [PROG] Phi) .
ceq Q |= [P(N,M)?] Phi = Q' |= Phi if Q' := (Q).P(N,M) .
```

where **var** and **vars** keywords are used to declare variables of some sorts. The first two equations define the semantics of the atomic formulas $p(i, |\psi\rangle)$ and $p(i, i + 1, |\Psi\rangle)$, where $i$ denotes the index at which the projections will take place and $|\psi\rangle, |\Psi\rangle$ are used to construct their projection operators in the forms of $|\psi\rangle\langle\psi|, |\Psi\rangle\langle\Psi|$, respectively. Recall that $(M, |\psi\rangle) \models p(i, |\psi\rangle)$ if and only if $P_{[\![p(i,|\psi\rangle)]\!]^M}(|\psi\rangle) = |\psi\rangle$. The next two equations define the negation of atomic formulas $p(i, |0\rangle)$ and $p(i, |1\rangle)$. It is not necessary to implement the negation of the other formulas for conducting the experiments in Sect. 5. The next equation reflects the semantics of conjunction. Based on Theorem 2, the remaining equations simulate **skip**, **abort**, quantum gates ($I$, $H$, $X$, $Y$, $Z$, and CX), ; (composition), $\cup$ (choice), and ? (test). Note that $A$? is implemented only for $A = p(i, |\psi\rangle)$ because the other more complex test operators are not needed for conducting the experiments in Sect. 5. For the sake of simplicity, we do not mention how we implement the behavior of quantum gates and projections in detail to make the paper concise.

Let us suppose that $E_{\text{SR}}$ and $E_{\text{BDQL}}$ are the sets of equations used for symbolic reasoning on quantum computation adopted from [11] and the semantics of BDQL specified in Maude, respectively. Now we have enough facilities to check whether $(M, s) \models A$ by simplifying s |= A $\rightarrow^*_{E_{\text{SR}} \cup E_{\text{BDQL}}}$ emptyJS. Indeed, $(M, s) \models A$ if s |= A is simplified to emptyJS with respect to $E_{\text{SR}} \cup E_{\text{BDQL}}$.

**Table 1.** Experimental results with our support tool for the five case studies

| Protocol | Qubits | Rewrite Steps | Time |
|---|---|---|---|
| Superdense Coding | 2 | 2,659 | 1 ms |
| Quantum Teleportation | 3 | 2,558 | 1 ms |
| Quantum Secret Sharing | 4 | 7,139 | 3 ms |
| Entanglement Swapping | 4 | 5,344 | 2 ms |
| Quantum Gate Teleportation | 6 | 56,901 | 37 ms |

## 5   Experiments

This section shows how to use our support tool to verify Quantum Teleportation in Maude as an example and similarly for other protocols, which can be fully found at https://github.com/canhminhdo/DQL. Subsequently, we provide the experimental results for five protocols used in the experiments.

Let TELEPORT be the specification of Quantum Teleportation, initQState be the initial state for TELEPORT and qubitAt be the function to get a single qubit at some index. We can verify the correctness of Quantum Teleportation with our support tool using the **reduce** command in Maude as follows:

```
reduce in TELEPORT : initQState |= [
  H(1) ; CX(1, 2) ; CX(0, 1) ; H(0) ;
  if P(1, |0>) then skip else X(2) fi ;
  if P(0, |0>) then skip else Z(2) fi
] P(2, qubitAt(initQState, 0)) .
```

The **reduce** command will conduct the simplification process with respect to the equations specified in our tool automatically. The command returns emptyJS in just a few moments, and thus the correctness of Quantum Teleportation is verified using our support tool, the implementation of BDQL in Maude, where the first qubit at index 0 of the initial quantum state is teleported correctly in the third qubit at index 2 of the final quantum state.

We conducted experiments on an iMac that carries a 4 GHz microprocessor with eight cores and 32 GB memory of RAM. The experimental results are shown in Table 1. We successfully verified the correctness of Superdense Coding, Quantum Teleportation, Quantum Secret Sharing, Entanglement Swapping, and Quantum Gate Teleportation according to the properties described in Sect. 3. For all case studies from two to six qubits used, we can quickly verify their correctness in just a few moments using our support tool, although the number of rewrite steps involved is quite large. Without the aid of computer programs, such as our support tool, these results would have been almost impossible. This demonstrates the effectiveness of our automated approach for verifying quantum programs in BDQL using the symbolic approach adopted from [11].

# 6    Related Work

Quantum Hoare Logic (QHL) by [25] was designed and intended to be a quantum counterpart of Hoare Logic. From the perspective of logic, BDQL can express more fundamental components of quantum programs compared to QHL: the **if** ⋯ **fi** statement that represents a non-deterministic measurement cannot be divided anymore in QHL. On the other hand, BDQL can express its non-deterministic feature explicitly using the choice operator ∪. Also, QHL lacks the test operator in its syntax.

In this paper, we chose Maude as our implementation language. On the other hand, [13] was implemented in PRISM (Probabilistic symbolic model checker) for verifying quantum protocols. Unlike our approach, [13] needed to enumerate states and calculate the state transitions in advance and then encode them into a specification. In contrast, our approach did not require such enumeration of states in advance because we formalized the quantum computation and the semantics of BDQL by means of equations, and the verification problem is conducted automatically through an equational simplification process in Maude.

# 7    Conclusions and Future Work

We have presented the implementation of BDQL, a simplified version of DQL, in Maude for quantum program verification. The symbolic reasoning from [11] is adopted, and we have formalized the semantics of BDQL by means of equations. The verification problem is simplified using the `reduce` command in Maude with respect to the equations specified in our support tool. Using our support tool, we have successfully verified the five quantum programs. This demonstrates the effectiveness of our automated approach for verifying quantum programs in BDQL using symbolic reasoning.

At least two future extensions remain to be addressed. That is, our tool (and BDQL) is limited in that (I) it cannot deal with programs including the Kleene star operator (iteration operator) and that (II) it cannot deal with quantitative properties regarding measurement probability. As to (I), it is significant to extend our tool so that it can deal with the Kleene star operator for expressing quantum loop programs [26]. As to (II), a DQL called the Probabilistic Logic of Quantum Programs (PLQP) that can express the quantitative properties has been proposed and applied to formal verification of the quantum search algorithm, Quantum Leader Election, and the BB84 quantum key distribution protocol [3,7]. However, their verification was done manually, and their automation by a tool is still an open problem.

**Acknowledgements.** The authors are grateful to the anonymous reviewers for their valuable feedback. The research was supported by JAIST Research Grant for Fundamental Research. The research of the first author was supported by Grant-in-Aid for JSPS Research Fellow Grant Number JP22KJ1483. The research of the second and the third authors was supported by JST SICORP Grant Number JPMJSC20C2 and JSPS KAKENHI Grant Number JP24H03370.

# Appendix

## Proof of Theorem 1

Before embarking on the proof of Theorem 1, we show a lemma.

**Lemma 1.** *The following holds for any $M$:*

1. $[\![[\mathbf{skip}]A]\!]^M = [\![A]\!]^M;$
2. $[\![[\mathbf{abort}]A]\!]^M = \mathcal{H};$
3. $[\![[\mathbf{abort};b]A]\!]^M = [\![[\mathbf{abort}]A]\!]^M;$
4. $[\![[\mathbf{skip};b]A]\!]^M = [\![[b]A]\!]^M;$
5. $[\![[(a\;;\;b)\;;\;c]A]\!]^M = [\![[a\;;\;(b\;;\;c)]A]\!]^M;$
6. $[\![[(a\cup b)\;;\;c]A]\!]^M = [\![[(a\;;\;c)\cup(b\;;\;c)]A]\!]^M;$
7. $[\![[a\;;\;b]A]\!]^M = [\![[a][b]A]\!]^M;$
8. $[\![[a\cup b]A]\!]^M = [\![[a]A]\!]^M \cap [\![[b]A]\!]^M;$
9. $[\![[B?]A]\!]^M = [\![B\to A]\!]^M \in \mathcal{C}(\mathcal{H}),$ where $B\to A$ denotes the Sasaki hook [18] defined as $\neg(A\wedge\neg(A\wedge B)).$

*Proof.* 1 to 8 are easy to show. For 9, some knowledge of Hilbert space theory is required. Observe that $[\![[B?]A]\!]^M$ is the inverse image $P_{[\![B]\!]^M}^{-1}([\![A]\!]^M)$ of $[\![A]\!]^M$ under $P_{[\![B]\!]^M}$. That is,

$$[\![[B?]A]\!]^M = P_{[\![B]\!]^M}^{-1}([\![A]\!]^M) = \{s\in\mathcal{H}:P_{[\![B]\!]^M}(s)\in[\![A]\!]^M\}$$
$$= \{s\in\mathcal{H}:P_{[\![A]\!]^M}P_{[\![B]\!]^M}(s) = P_{[\![B]\!]^M}(s)\}.$$

Therefore, $[\![[B?]A]\!]^M = [\![B\to A]\!]^M \in \mathcal{C}(\mathcal{H})$ (see [15]).

We use Lemma 1 to prove Theorem 1 without mentioning it.

*Proof.* We prove by simultaneous structural induction on formulas in BDQL and star-free regular programs. The case $A = p \in L_0$ is immediate. The cases $A = \neg B$ and $A = B\wedge C$ follow from the basic fact in Hilbert space theory. Thus, we only discuss the case $A = [a]B$.

**Case 1** $a = \mathbf{skip}$. We have $[\![[a]B]\!]^M = [\![B]\!]^M \in \mathcal{C}(\mathcal{H})$ by the induction hypothesis $[\![B]\!]^M \in \mathcal{C}(\mathcal{H})$.

**Case 2** $a = \mathbf{abort}$. We have $[\![[a]B]\!]^M = \mathcal{H} \in \mathcal{C}(\mathcal{H})$.

**Case 3** $a = \pi \in \Pi_0$. Observe that $[\![[a]B]\!]^M$ is the inverse image of $[\![B]\!]^M$ under $v(a)$. In other words, $[\![[a]B]\!]^M$ is the image $(v(a)^\dagger)([\![B]\!]^M)$ of $[\![B]\!]^M$ under the adjoint operator $v(a)^\dagger$ of $v(a)$. Let $X^\perp$ be the orthogonal complement of a subspace $X$ of $\mathcal{H}$, and write $X^{\perp\perp}$ for $(X^\perp)^\perp$. Recall that $X \in \mathcal{C}(\mathcal{H})$ if and only if $X^{\perp\perp} = X$. By the induction hypothesis $[\![B]\!]^M \in \mathcal{C}(\mathcal{H})$,

$$([\![[a]B]\!]^M)^{\perp\perp} = ((v(a)^\dagger)([\![B]\!]^M))^{\perp\perp} = ((v(a)^\dagger)(([\![B]\!]^M)^\perp))^\perp$$
$$= (v(a)^\dagger)(([\![B]\!]^M)^{\perp\perp}) = (v(a)^\dagger)([\![B]\!]^M) = [\![[a]B]\!]^M.$$

Consequently, $[\![[a]B]\!]^M \in \mathcal{C}(\mathcal{H})$.

**Case 4** $a = b \; ; \; c$. We further split the case with respect to $b$.

**Case 4.1** $b = \textbf{skip}$. $[\![a]B]\!]^M = [\![c]B]\!]^M \in \mathcal{C}(\mathcal{H})$ by the induction hypothesis $[\![c]B]\!]^M \in \mathcal{C}(\mathcal{H})$.

**Case 4.2** $b = \textbf{abort}$. $[\![a]B]\!]^M = [\![\textbf{abort}]B]\!]^M = \mathcal{H} \in \mathcal{C}(\mathcal{H})$.

**Case 4.3** $b = \pi$. $[\![a]B]\!]^M = [\![\pi][c]B]\!]^M$. By the induction hypothesis, $[\![c]B]\!]^M \in \mathcal{C}(\mathcal{H})$. Thus, it follows from the similar argument of case 3 above that $[\![a]B]\!]^M \in \mathcal{C}(\mathcal{H})$.

**Case 4.4** $b = b_1 \; ; \; b_2$.

$$[\![a]B]\!]^M = [\![b_1 \; ; \; (b_2 \; ; \; c)]B]\!]^M = [\![b_1][b_2 \; ; \; c]B]\!]^M \in \mathcal{C}(\mathcal{H})$$

by the induction hypothesis $[\![b_1][b_2 \; ; \; c]B]\!]^M \in \mathcal{C}(\mathcal{H})$.

**Case 4.5** $b = b_1 \cup b_2$.

$$[\![a]B]\!]^M = [\![(b_1 \; ; \; c) \cup (b_2 \; ; \; c)]B]\!]^M = [\![b_1 \; ; \; c]B]\!]^M \cap [\![b_2 \; ; \; c]B]\!]^M \in \mathcal{C}(\mathcal{H})$$

by the induction hypothesis $[\![b_1 \; ; \; c]B]\!]^M, [\![b_2 \; ; \; c]B]\!]^M \in \mathcal{C}(\mathcal{H})$.

**Case 4.6** $b = C?$.

$$[\![a]B]\!]^M = [\![C?][c]B]\!]^M = [\![C \to [c]B]\!]^M \in \mathcal{C}(\mathcal{H})$$

by the induction hypothesis $[\![C]\!]^M, [\![c]B]\!]^M \in \mathcal{C}(\mathcal{H})$.

**Case 5** $a = b \cup c$. We have $[\![a]B]\!]^M = [\![b]B]\!]^M \cap [\![c]B]\!]^M \in \mathcal{C}(\mathcal{H})$ by the induction hypothesis $[\![b]B]\!]^M, [\![c]B]\!]^M \in \mathcal{C}(\mathcal{H})$.

**Case 6** $a = C?$. We have $[\![C?]B]\!]^M = [\![C \to B]\!]^M \in \mathcal{C}(\mathcal{H})$ by the induction hypothesis $[\![B]\!]^M, [\![C]\!]^M \in \mathcal{C}(\mathcal{H})$.

# References

1. Akatov, D.: The logic of quantum program verification. Master's thesis, Oxford University (2005)
2. Baltag, A., Smets, S.: LQP: the dynamic logic of quantum information. Math. Struct. Comput. Sci. **16**(3), 491–525 (2006)
3. Baltag, A., Bergfeld, J., Kishida, K., Sack, J., Smets, S., Zhong, S.: PLQP & company: decidable logics for quantum algorithms. Int. J. Theor. Phys. **53**(10), 3628–3647 (2014)
4. Baltag, A., Smets, S.: Reasoning about quantum information: an overview of quantum dynamic logic. Appli. Sci. **12**(9) (2022)
5. Bennett, C.H., Brassard, G., Crépeau, C., Jozsa, R., Peres, A., Wootters, W.K.: Teleporting an unknown quantum state via dual classical and Einstein-Podolsky-Rosen channels. Phys. Rev. Lett. **70**, 1895–1899 (1993)
6. Bennett, C.H., Wiesner, S.J.: Communication via one- and two-particle operators on Einstein-Podolsky-Rosen states. Phys. Rev. Lett. **69**, 2881–2884 (1992)
7. Bergfeld, J.M., Sack, J.: Deriving the correctness of quantum protocols in the probabilistic logic for quantum programs. Soft. Comput. **21**(6), 1421–1441 (2017)
8. Biamonte, J., Wittek, P., Pancotti, N., Rebentrost, P., Wiebe, N., Lloyd, S.: Quantum machine learning. Nature **549**(7671), 195–202 (2017)
9. Birkhoff, G., von Neumann, J.: The logic of quantum mechanics. Ann. Math. **57**(4), 823–843 (1936)

10. Clavel, M., Durán, F., Eker, S., Lincoln, P., Martí-Oliet, N., Meseguer, J., Talcott, C.L. (eds.): All About Maude - A High-Performance Logical Framework, How to Specify, Program and Verify Systems in Rewriting Logic. LNCS, vol. 4350. Springer (2007). https://doi.org/10.1007/978-3-540-71999-1

11. Do, C.M., Ogata, K.: Symbolic model checking quantum circuits in maude. In: The 35th International Conference on Software Engineering and Knowledge Engineering, SEKE 2023 (2023)

12. Farhi, E., Goldstone, J., Gutmann, S., Lapan, J., Lundgren, A., Preda, D.: A quantum adiabatic evolution algorithm applied to random instances of an np-complete problem. Science **292**(5516), 472–475 (2001)

13. Gay, S., Nagarajan, R., Papanikolaou, N.: Probabilistic model-checking of quantum protocols. arXiv preprint quant-ph/0504007 (2005)

14. Gottesman, D., Chuang, I.L.: Demonstrating the viability of universal quantum computation using teleportation and single-qubit operations. Nature **402**(6760), 390–393 (1999)

15. Hardegree, G.M.: The conditional in quantum logic. Synthese, 63–80 (1974)

16. Harel, D., Kozen, D., Tiuryn, J.: Dynamic Logic. MIT Press (2000)

17. Harrow, A.W., Hassidim, A., Lloyd, S.: Quantum algorithm for linear systems of equations. Phys. Rev. Lett. **103**, 150502 (2009)

18. Herman, L., Marsden, E.L., Piziak, R.: Implication connectives in orthomodular lattices. Notre Dame J. Formal Logic **16**(3), 305–328 (1975)

19. Hillery, M., Bužek, V., Berthiaume, A.: Quantum secret sharing. Phys. Rev. A **59**, 1829–1834 (1999)

20. Meseguer, J.: Twenty years of rewriting logic. J. Log. Algebraic Methods Program **81**(7–8), 721–781 (2012)

21. Nielsen, M.A., Chuang, I.L.: Quantum Computation and Quantum Information: 10th Anniversary Edition. Cambridge University Press (2011)

22. Paykin, J., Rand, R., Zdancewic, S.: Qwire: a core language for quantum circuits. SIGPLAN Not. **52**(1), 846–858 (2017)

23. Rédei, M.: Quantum logic in algebraic approach. FTPH, vol. 91. Springer (1998). https://doi.org/10.1007/978-94-015-9026-6

24. Shor, P.: Algorithms for quantum computation: discrete logarithms and factoring. In: Proceedings 35th Annual Symposium on Foundations of Computer Science, pp. 124–134 (1994)

25. Ying, M.: Floyd-hoare logic for quantum programs. ACM Trans. Program. Lang. Syst. (TOPLAS) **33**(6), 1–49 (2012)

26. Ying, M., Feng, Y.: Quantum loop programs. Acta Informatica **47**(4), 221–250 (2010)

27. Ying, M., Feng, Y.: Model checking quantum systems – a survey (2018)

28. Żukowski, M., Zeilinger, A., Horne, M.A., Ekert, A.K.: "Event-ready-detectors" Bell experiment via entanglement swapping. Phys. Rev. Lett. **71**, 4287–4290 (1993)

# Predictive Theory of Mind Models Based on Public Announcement Logic

Jakob Dirk Top[1(✉)], Catholijn Jonker[2,3], Rineke Verbrugge[1], and Harmen de Weerd[1]

[1] University of Groningen, Groningen, The Netherlands
{j.d.top,l.c.verbrugge,harmen.de.weerd}@rug.nl
[2] Delft University of Technology, Delft, The Netherlands
c.m.jonker@tudelft.nl
[3] Leiden University, Leiden, The Netherlands

**Abstract.** Epistemic logic can be used to reason about statements such as 'I know that you know that I know that $\varphi$'. In this logic, and its extensions, it is commonly assumed that agents can reason about epistemic statements of arbitrary nesting depth. In contrast, empirical findings on Theory of Mind, the ability to (recursively) reason about mental states of others, show that human recursive reasoning capability has an upper bound.

In the present paper we work towards resolving this disparity by proposing some elements of a logic of *bounded* Theory of Mind, built on Public Announcement Logic. Using this logic, and a statistical method called Random-Effects Bayesian Model Selection, we estimate the distribution of Theory of Mind levels in the participant population of a previous behavioral experiment. Despite not modeling stochastic behavior, we find that approximately three-quarters of participants' decisions can be described using Theory of Mind. In contrast to previous empirical research, our models estimate the majority of participants to be second-order Theory of Mind users.

**Keywords:** Theory of Mind · Public Announcement Logic · Epistemic Logic · Behavioral Modeling · Random-Effects Bayesian Model Selection · Cognitive Science

## 1 Introduction

Theory of Mind (ToM) is the ability to attribute and reason about mental states of others, such as knowledge, beliefs, and intentions [10,30]. ToM can be used recursively. For example, if Amy knows that Ben knows that Amy knows that there will be a surprise party, Amy is using second-order ToM (ToM-2), by reasoning about the way Ben is using his theory of mind to reason about her own knowledge; and we are making a third-order attribution to Amy here. ToM is commonly used to navigate social situations, and can improve the outcomes of

N. Gierasimczuk and F. R. Velázquez-Quesada (Eds.): DaLí 2023, LNCS 14401, pp. 85–103, 2024.
https://doi.org/10.1007/978-3-031-51777-8_6

competitive [16,32], cooperative [13,28], and mixed-motive settings [39]. While human ToM capabilities develop over early childhood [41], and can be trained [1,38,40], it is generally found that there is a limit to human recursive ToM use, which often does not exceed level 2 [7,9,12,27], and sometimes fails entirely [23].

Epistemic logic, a variant of modal logic, is used to formalize the kind of recursive knowledge needed for ToM statements of the form 'I know that you know...' [19]. However, epistemic logics and their extensions classically assume logical omniscience, contrary to the commonly found limits on ToM. It has been suggested that these models should incorporate recursive reasoning limits [17,39], and there have been previous attempts to model similar aspects of bounded rationality [8,11,24,31]. The first formal attempt to incorporate ToM-like limitations in epistemic logic appears to be [22], which describes an approach close to our purposes: They define the *epistemic depth* of a formula based on the nesting of its modal operators. However, their approach does not cover Public Announcement Logic (PAL, introduced in Sect. 2.2), which we require for our purposes, and is a general approach that does not define how it can be used to encode the specific attributes of ToM.

While formal methods often do not take into account the ToM limits found in behavioral research, the latter does not regularly employ the tasks and models commonly used in epistemic logic, such as epistemic puzzles. Epistemic puzzles, like the Wise Men puzzle [25], Muddy Children puzzle [14], and the one described in Sect. 2.1, are puzzles where a set of agents, in a partially observable world, have to deduce unobservable facts using the epistemic statements of other agents. In the literature, reproducible experiments using these puzzles, especially ones yielding reusable data, appear sparse (see e.g. [8,18,20]).

The present paper attempts to bridge the gap between logic and (boundedly rational) cognition. We build on the work of Cedegao and colleagues [8] by adding ToM limitations to PAL, which we use to predict the answers of different ToM levels in the game of Aces and Eights (explained in Sect. 2.1). We validate our novel method on the data of Cedegao et al. [8] by using Random-Effects Bayesian Model Selection (RFX-BMS) [33], which we use to estimate the frequencies of different ToM levels among the participants of [8].

In recent work, parallel to ours, Arthaud and Rinard [2] create several logics of public announcements which place a limit on the number of nested knowledge operators an agent can understand. Before we continue, we note some key differences with our work. In [2], any nested knowledge operator increases a formula's depth, whereas we assume that only switching between knowledge operators for *different* agents requires higher ToM [39]. There should be a quantitative difference between recursively reasoning about your own knowledge, and that of others. In [2], a formula $K_a\varphi$ is false if the depth of agent $a$ is lower than that of $\varphi$. Our ToM-0 agents act as if there are no relations for other agents. If an agent has no outgoing relations, it vacuously knows everything, so ToM-0 agents know that all other agents know everything. This could be similar to young children without ToM, who may think that their parents are all-knowing [5]. Lastly, we move beyond purely formal methods by fitting our models on human data.

In Sect. 2, we explain the tasks, data, and methods we use for predictive modeling. In Sect. 3, we present the results of our novel predictive modeling

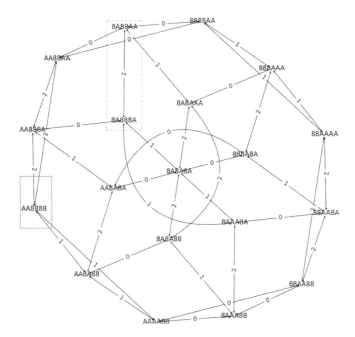

**Fig. 1.** Model before announcements. Reflexive edges omitted for clarity.

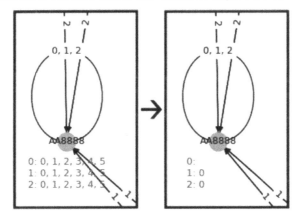

**Fig. 2.** State $AA8888$ in the ToM model before and after player 0 announces 'I do not know my cards'. This is a close-up of the orange rectangle in Fig. 1, with added ToM levels. Refer to Sect. 2.4 for an in-depth explanation.

method, and compare it to the results we obtain when applying Random-Effects Bayesian Model Selection to the models of [8]. Lastly, in Sect. 4, we discuss our findings and identify possible shortcomings and directions for future work.

# 2  Methods

Here, Sects. 2.1 through 2.3 describe existing work, leading into our novel work as described in Sects. 2.4 and 2.5.

## 2.1  The Game of Aces and Eights

Aces and Eights [14] is a three-player epistemic game where each player receives two cards out of a deck of four Aces and four Eights. Each player can only see the four cards held by the other two players. No player can see her own cards or the two remaining cards. Players take turns, in a fixed order, announcing whether or not they know the *ranks* of the cards they are holding — a card's suit does not matter. These announcements provide information that may allow players to work out which cards they have. Players are collectively informed of all these rules, allowing common knowledge of the game rules to arise.[1]

Let us introduce the notation employed throughout this paper. We use 'player 0', 'player 1', and 'player 2' (or, in short, '0', '1' and '2') for the player that makes the first, second, and third announcement each round, respectively. Suppose 0 has two aces ($AA$), 1 has two eights (88), and 2 has two eights (88). We denote the state of this game as $AA8888$, where the first two symbols are 0's cards, the second two symbols are 1's cards, and the third two symbols are 2's cards. In this state, 0 knows her cards. She sees that all available Eights are held by the other two players, so she must have two Aces. After 0 announces 'I know my cards', 1 and 2 can also know their cards, because they can attribute this reasoning to 0. For holding one Ace and one Eight (or one Eight and one Ace, as order does not matter), we write '$8A$'.

Cedegao et al. [8] discuss an experiment where each of 306 participants played ten games of Aces and Eights with two computer players that are perfect logical reasoners. Participants were recruited and played online, on the Prolific platform. The order and selection of games varied across participants, but each participant played one game requiring epistemic level 0 (EL-0, see Sect. 2.3) to solve, three games requiring EL-1, and two games each requiring EL-2, EL-3, and EL-4 (retrieved from their code). Participants switched between playing as player 0, 1, and 2 across games. Participants knew the rules and knew that the computer agents gave perfect answers. A game ended if the participant answered 'I know my cards', if the participant answered incorrectly (including answering 'I don't know' when they could have known), or if playing more rounds would not provide more information. Participants responding with 'I know my cards' also had to state the cards they thought they had. Participants were paid $5 with a $0.50 bonus for each game correctly solved. Participants were excluded if they failed more than 20% of attention checks, spent more than 87 min, gave impossible responses according to the rules, or had data recording errors. Following [8], this paper only uses the data for the remaining 211 participants.

---

[1] For solving the game of Aces and Eights, all players also need to be truthful, perfect logical reasoners, and there needs to be common knowledge of this.

## 2.2   Public Announcement Logic

Public Announcement Logic (PAL) [3,4,29] is an extension of epistemic logic that models how the knowledge of agents changes after public announcements are made. Here, the knowledge of all agents in some epistemic situation is encoded in a *Kripke model* (thus, assuming logical omniscience). A Kripke model can be represented using a directed graph. The graph for Aces and Eights is found in Fig. 1. Each node, or state, is a possible situation, such as the distribution of cards in Aces and Eights. Each edge is labelled with player(s), and indicates uncertainty for those players: A player $i$ edge from state $s_1$ to $s_2$ means 'if $s_1$ is the *true* state (the state corresponding to the actual distribution of cards), then player $i$ considers it possible that $s_2$ is the true state' (here, we may have $s_1 = s_2$). For example, if 2 sees that 0 has $8A$ and 1 has 88, then 2 considers it possible that she has either $8A$ or $AA$, so there is a symmetric player 2 edge between $8A888A$ and $8A88AA$, as well as reflexive edges at both states. This situation can be found in Fig. 1, where it is indicated with a cyan, dashed, rectangle (reflexive edges omitted). If, in state $s$, all outgoing player $i$ edges connect to worlds where $i$ has the same cards, then $i$ knows her cards. An example of this is player 0 in $AA8888$, found in the solid orange rectangle in Fig. 1.

## 2.3   Bounded Models

Cedegao et al. [8] model an epistemic level $l$ as follows: Take as an agent's *initial states* those states that the agent considers possible based on the game rules and the cards held by the other two players. For example, if agent 1 sees that 0 holds $AA$ and 2 holds $8A$, then agent 1's initial states are $AA888A$ and $AA8A8A$. Modifying Definition 2.32 of [6], the height of a state is defined by induction: the height of *all* initial states is 0, and the states of height $n+1$ are the immediate successors (states that can be reached in one step along any outgoing edge) of states of height $n$ that have not yet been assigned a height smaller than $n+1$. States with height $l$ are marked peripheral states, and their outgoing edges are removed. States with a height exceeding $l$ are removed entirely. When an announcement is made, a bounded model is updated by removing those *non-peripheral* states (and connecting edges) where the announced formula is false. Answers are based on the remaining initial states. Since our models differ from those in [8], we use 'ToM order' when talking about our models, and 'epistemic level' (EL) when talking about the models of [8].

Since all states *other than* the peripheral states have the same relations as the full model, which is an $S5_{(3)}$-model, Cedegao's models allow for paths with an infinite number of switches between *different* agents (e.g., $\ldots (s_{n-1}, s_n) \in R(0), (s_n, s_{n+1}) \in R(1), (s_{n+1}, s_{n+2}) \in R(0), \ldots$). We argue that paths with infinitely many perspective switches are contrary to human recursive ToM limits. Furthermore, an agent with epistemic level 4, playing Aces and Eights, uses the same graph as a logically omniscient agent. In contrast to Cedegao and colleagues [8], we instead attempt to limit the number of recursive reasoning steps an agent can use, as outlined in the next section.

## 2.4   Theory of Mind Models

This section introduces our novel methods for modeling ToM, in a logic we call TOMPAL. We work in the language $\mathcal{L}_{K[]}(A, P)$, taken directly from [37]:

**Definition 1.** The language of public announcement logic is inductively defined

$$\mathcal{L}_{K[]}(A, P) \ni \varphi ::= \quad p \quad | \quad \neg\varphi \quad | \quad (\varphi \wedge \varphi) \quad | \quad K_i\varphi \quad | \quad [\varphi]\varphi$$

with $i \in A$, a set of agents, and $p \in P$, a finite set of propositional atoms.

The usual abbreviations are used for $\vee$, $\rightarrow$, and $\leftrightarrow$. For $\neg K_i \neg \varphi$ we use $M_i\varphi$.

We consider that repeated nestings of knowledge operators for the same agent do not require additional ToM levels to be understood (see [39]), and that reasoning about one's own knowledge does not require ToM at all. Instead, we assume only *switching* to the perspective of a different agent requires an additional level of ToM. For example, player 0 needs ToM-2 to reason about the sentence $K_0 K_0 K_1 K_1 K_1 K_0 p$.[2] When an agent switches perspectives, she attributes her own order, minus one, to the other agent. To keep track of this, we modify the definition of models in [36] by adding a map $T$, as follows:

**Definition 2.** A ToM model $M = (S, R, V, T)$ consists of a non-empty set of states $S$, an accessibility function $R : A \rightarrow \mathcal{P}(S \times S)$, a valuation $V : P \rightarrow \mathcal{P}(S)$, where $V(p)$ is the set of states where $p$ is true, and a ToM map $T : S \rightarrow \mathcal{P}(A \times \mathbb{N})$ (with $0 \in \mathbb{N}$), which maps each state to a set of tuples $(i, l)$ with $i \in A$ and $l \in \mathbb{N}$. For $s \in S$, $i \in A$, and $l \in \mathbb{Z}$, the pair $(M, (s, (i, l)))$ is a **perspective state**.

Intuitively, having $(i, l) \in T(s)$ means 'agent $i$, at ToM order $l$, has not yet eliminated state $s$ due to new information'. Conversely, $(i, l) \notin T(s)$ means 'agent $i$, at ToM order $l$, either due to some previous announcement no longer considers state $s$ to be possible, or did not consider it possible to begin with'.

Visually, to each state in the model found in Fig. 1 we add one row for each player, consisting of the player's name, followed by a colon, followed by

---

[2] Note that this differs from [11], where the *horizon* of a player $i$ at $(M, s)$ contains all states player $i$ can 'reach' by taking one step along one of her own edges, followed by any number of steps along any agent's edges. Closer to our intentions, but more general, is the notion of *admissibility on E* [22,24].

that player's possible ToM levels, e.g., '0: 0, 1, 2, 3, 4, 5' at state $s$ means $(0,0) \in T(s), (0,1) \in T(s), \ldots, (0,5) \in T(s)$. An example for state $AA8888$ can be found in the leftmost half of Fig. 2. Here, considering it possible that the actual distribution of cards is $AA8888$ is consistent with reasoning at ToM levels 0 through 5 for all players. In our software implementation of Aces and Eights, we ignore ToM levels beyond 5, because these yield identical answers to ToM-5.

A *perspective state* is an epistemic state viewed from the perspective of agent $i$ at ToM order $l$; such states are used in our semantics. The semantics of TOMPAL are a modification of those in [37] and are as follows:

**Definition 3.** Assuming a ToM model $M = (S, R, V, T)$, $i \in A$, and $l \in \mathbb{Z}$:

$$
\begin{aligned}
M, (s, (i, l)) &\models p & &\Leftrightarrow s \in V(p) \\
M, (s, (i, l)) &\models \neg\varphi & &\Leftrightarrow M, (s, (i, l)) \not\models \varphi \\
M, (s, (i, l)) &\models \varphi \wedge \psi & &\Leftrightarrow M, (s, (i, l)) \models \varphi \text{ and } M, (s, (i, l)) \models \psi \\
\text{for } i = j : M, (s, (i, l)) &\models K_j\varphi & &\Leftrightarrow M, (t, (j, l)) \models \varphi \text{ for all } (t, (j, l)) \text{ with} \\
& & &\qquad (s, t) \in R(j) \text{ and } (j, l) \in T(t) \\
\text{for } i \neq j : M, (s, (i, l)) &\models K_j\varphi & &\Leftrightarrow M, (t, (j, l-1)) \models \varphi \text{ for all } (t, (j, l-1)) \text{ with} \\
& & &\qquad (s, t) \in R(j) \text{ and } (j, l-1) \in T(t) \\
M, (s, (i, l)) &\models [\varphi]\psi & &\Leftrightarrow M, (s, (i, l)) \models \varphi \text{ implies } M|\varphi, (s, (i, l)) \models \psi
\end{aligned}
$$

where the model restriction $M|\varphi = (S, R, V, T')$ is defined as $(i, l) \in T'(s)$ iff $(i, l) \in T(s)$ and $[M, (s, (i, l)) \models \varphi$ or $[l \leq 0$ and $\varphi$ contains an operator $K_j$ with $i \neq j]]$.

We make three deviations from the usual semantics for public announcement logic: first, formulas are interpreted at a perspective state $M, (s, (i, l))$. They are true or false from the perspective of a specific agent with a specific ToM order. Secondly, our knowledge operator has two clauses: when an agent reasons about her own knowledge, she does not switch perspectives. When an agent reasons about the knowledge of a different agent, she switches perspectives to the other agent, and attributes her own ToM order, minus one, to the other agent. In doing so, a ToM-0 agent attributes ToM-(-1) to other agents. Since by definition we have $(i, -1) \notin T(s)$ for all $i$ and $s$, a ToM-0 agent reasons as if there are no outgoing relations for other agents. Lastly, we modify the model restriction such that tuples $(i, l)$ are removed instead of states. A ToM-0 agent cannot switch perspectives, and therefore 'ignores' announcements that she cannot understand because they contain $K$-operators for other agents.[3]

Next, we show some theorems that capture the properties of TOMPAL. First, we want ToM-0 agents to ignore announcements they do not understand. From a ToM-0 agent's perspective, no tuples are removed due to such announcements:

---

[3] We use $l = 0$ as the only special case, but for situations other than Aces and Eights we need a more general solution, found in Appendix A. Furthermore, our semantics can be made equivalent to one with the usual knowledge operator if we 'unfold' our models such that we have $R : (A \times \mathbb{N}) \to \mathcal{P}(S \times S)$.

**Theorem 1.** *If $\varphi$ contains a $K_j$ operator, then for all $M, (s, (i, 0))$ with $i \neq j$:*

$$M, (s, (i, 0)) \models (\varphi \rightarrow \psi) \leftrightarrow [\varphi]\psi.$$

*Proof.* The key point is showing that $T' = T$ and hence $M|\varphi = M$. Details are left to the reader. □

Secondly, ToM-0 agents should act as if there are no outgoing relations for *other* agents, so we should have:

**Theorem 2.** *For all $M, (s, (i, 0))$ with $i \neq j$: $M, (s, (i, 0)) \models K_j\varphi$.*

*Proof.* The key point is that there are no $(t, (j, -1))$ with $(s, t) \in R(j)$ and $(j, -1) \in T(s)$, due to the definition of $T$. Details are left to the reader. □

Note that Theorem 2 implies that $M, (s, (i, 0)) \models K_j\varphi \wedge K_j\neg\varphi$ when $i \neq j$.

Lastly, there should be no paths which infinitely alternate between *different* agents, as ToM puts a limit on the number of times any agent can switch perspectives:

**Theorem 3.** *For all non-empty sequences $(M_{j_1} M_{j_2}, \ldots, M_{j_{n-1}}, M_{j_n})$ of M-operators such that $|\{k : j_k \neq j_{k+1}\}| > l \geq 0$, respectively for all $M, (s, (i, l))$ and for all $M, (s, (i, l+1))$:*

*Clause 1:* $M, (s, (i, l)) \quad \models \neg M_{j_1} M_{j_2} \ldots M_{j_{n-1}} M_{j_n} \psi \qquad$ *for $i = j_1$*
*Clause 2:* $M, (s, (i, l+1)) \models \neg M_{j_1} M_{j_2} \ldots M_{j_{n-1}} M_{j_n} \psi \qquad$ *for $i \neq j_1$*

*Proof.* First, we denote $M_{j_1} M_{j_2} \ldots M_{j_{n-1}} M_{j_n}$ as $M^n$. We rewrite $\neg M^n \psi$ as $K^n \neg\psi$, which, as we prove for all $\psi \in \mathcal{L}_{K[]}$, we rewrite to $K^n \psi$. We prove the theorem through mutual induction over $l$.

**Base case, clause 2:** our base case is that for all $M, (s, (i, 0))$ with $i \neq j_1$: $M, (s, (i, 0)) \models K_{j_1} \ldots K_{j_n} \psi$, which is shown in Theorem 2 by taking $K_{j_1}$ as $K_j$ and $K_{j_2} \ldots K_{j_n} \psi$ as $\varphi$.

**Inductive step from clause 2 to clause 1:** our **induction hypothesis** is that for some arbitrary $l \geq 0$, for all $M, s, i$ with $i \neq j_1$: $M, (s, (i, l)) \models K^n \psi$. We have to show that, for some non-empty sequence $(K_i, \ldots, K_i)$, $M, (s, (i, l)) \models K_i \ldots K_i K^n \psi$. For $s$ we write $s_1$, for $(K_i, \ldots, K_i)$ we write $(K_{i_1}, \ldots, K_{i_m})$. We omit all text after the first 'for all':

$M, (s_1, (i, l)) \quad \models K_{i_1} K_{i_2} \ldots K_{i_m} K^n \varphi \qquad\qquad\qquad\qquad \Leftrightarrow$
$M, (s_2, (i, l)) \quad \models K_{i_2} \ldots K_{i_m} K^n \varphi$ for all $(s_2, (i, l))$ with
$\qquad\qquad\qquad\quad (s_1, s_2) \in R(i)$ and $(i, l) \in T(s_2) \qquad\qquad \Leftrightarrow$

$\vdots \qquad\qquad\qquad \vdots \qquad\qquad\qquad\qquad\qquad\qquad\qquad \vdots$

$M, (s_m, (i, l)) \quad \models K_{i_m} K^n \varphi$ for all $\ldots \qquad\qquad\qquad\qquad \Leftrightarrow$
$M, (s_{m+1}, (i, l)) \models K^n \varphi$ for all $\ldots$

The latter holds because of our induction hypothesis.

**Inductive step from clause 1 to clause 2:** our **induction hypothesis** is $M, (s, (i, l)) \models K^n \psi$ for some arbitrary $M, (s, (i, l))$ with $l \geq 0$, and $i = j_1$.

We have to show that for $i \neq k$, $M, (s, (k, l + 1)) \models K^n\psi$. Through a series of equivalences, it can be shown that both are equivalent to $M, (t, (j_1, l)) \models K_{j_2} \dots K_n\psi$ for all $(t, (j_1, l))$ with $(s, t) \in R(j_1)$ and $(j_1, l) \in T(t)$.

By starting at our base case for clause 2 and *alternating* between both inductive steps, any instance of the theorem can be constructed. No base case for clause 1 is needed. For all $M, (s, (i, l))$ with $l < 0$, $K^n\psi$ holds vacuously, as by definition $(i, l) \notin T(s)$ and $(i, l - 1) \notin T(s)$. $\qquad\square$

**Aces and Eights.** For Aces and Eights, we use $A = \{0, 1, 2\}$ and $P = \{88_0, 8a_0, aa_0, 88_1, 8a_1, aa_1, 88_2, 8a_2, aa_2\}$, where $88_0$ means 'agent 0 is holding two eights', $8a_1$ means 'agent 1 is holding an Ace and an Eight', et cetera. $S$ and $R$ are as depicted in Fig. 1. $V$ is as would be expected. For example, $V(aa_0) \cap V(88_1) \cap V(88_2) = \{AA8888\}$. We have $(i, l) \in T(s)$ for all $s \in S$, $i \in A$, and $l \in \mathbb{N}$ (though we do not consider $l > 5$). Agent $i$ announcing 'I know my cards' is a public announcement of $K_i 88_i \vee K_i 8a_i \vee K_i aa_i$, announcing 'I do not know my cards' is a public announcement of its negation.

Consider state $AA8888$ in the leftmost half of Fig. 2, with for $AA8888$ only $(AA8888, AA8888) \in R(0)$. As an example, we show what happens to this state when agent 0 announces that she does *not* know her cards ($AA8888$ may not be the *true* state). For brevity, we use the simpler announcement 'I do not know that I have two Aces'. We compute $T'(AA8888)$ for $M|\neg K_0 aa_0$ (and hence $M|\neg K_0 aa_0$ itself). We consider each type of tuples on a case by case basis:

For tuples of the type $(i, 0)$ with $i \neq 0$, the formula contains an operator $K_j$ with $i \neq j$ and $l = 0$, so, by definition, these tuples are not removed.

For tuples of the type $(i, l)$ with $i \neq 0$ and $l > 0$, we have that $l \neq 0$, so we have to check whether $M, (AA8888, (i, l)) \models \neg K_0 aa_0$. If not, they are removed. We use a series of equivalences:

$M, (AA8888, (i, l)) \models \neg K_0 aa_0$ $\hspace{2cm}$ $\Leftrightarrow$ (definition of $\neg$)
$M, (AA8888, (i, l)) \not\models K_0 aa_0$ $\hspace{2cm}$ $\Leftrightarrow$ (def. of $K$)
$M, (t, (0, l - 1)) \not\models aa_0$ for some $(t, (0, l - 1))$ with
$\quad (AA8888, t) \in R(0)$ and $(0, l - 1) \in T(AA8888)$ $\hspace{0.8cm}$ $\Leftrightarrow$ (def. of $R(0)$)
$M, (AA8888, (0, l - 1)) \not\models aa_0$ for $(0, l - 1) \in T(AA8888)$.

We have $(0, 0), (0, 1), \dots, (0, 5) \in T(AA8888)$ and $AA8888 \in V(aa_0)$, so $M, (AA8888, (i, l)) \models \neg K_0 aa_0$ is false for any $i \neq 0$ and $l > 0$. Hence, all tuples of the type $(i, l)$ with $i \neq 0$ and $l > 0$ are removed. For similar reasons, all tuples of the type $(0, l)$ for *all* $l$ are also removed. The resulting $T'(AA8888)$ can be found in the rightmost half of Fig. 2.

**Answers.** With these TOMPAL models, we can model which answer any player $i$ with ToM level $l$ would give, given a distribution of cards (corresponding to state $s$) and a sequence of previous announcements, as follows: Using the methods previously described in this section, update the model with all previous announcements in order. Then, if exactly one of $M, (s, (i, l)) \models K_i 88_i$,

$M, (s, (i, l)) \models K_i 8a_i$, and $M, (s, (i, l)) \models K_i aa_i$ holds, player $i$ answers 'I know my cards', and states the cards she has. In any other case, player $i$ answers that she does not know her cards. Note that this deviates from standard epistemic logic where, if there are no outgoing edges for an agent $i$, all statements of the type 'agent $i$ knows $\varphi$' are true, whereas all statements 'agent $i$ does not know $\varphi$' are false. Recall from Sect. 1 that we will use TOMPAL to predict the answers and usage of different ToM levels in [8]'s data of Aces and Eights. To be able to employ our statistical methods, we need our models to give single answers. Not only is 'I do not know' the most common answer in the data, but it is also an intuitively good response when you consider nothing to be possible.

### 2.5 Random-Effects Bayesian Model Selection

Random-Effects Bayesian Model Selection (RFX-BMS) is a statistical method that estimates the frequencies of a set of strategies occurring in a population. Whereas fixed-effects Bayesian model selection methods assume there is a single strategy which best fits all participants, RFX-BMS assumes each subject was drawn from a fixed distribution of strategies, and estimates this distribution. Unlike Maximum Likelihood Estimation, RFX-BMS allows us to make more general claims about this distribution, and is robust to small differences between participants and strategies [9,33,38]. In our case, we estimate the frequencies of ToM levels in the participant population of Cedegao et al. [8]. RFX-BMS uses equation (14) of [33], which maximizes the log-likelihood of each participant using each ToM level by iteratively updating the strategy frequencies until convergence. This log-likelihood is $n(1 - \varepsilon) \cdot \ln(1 - \varepsilon) + n\varepsilon \cdot \ln(p \cdot \varepsilon)$, where a ToM level's error rate $\varepsilon$ for a participant is its number of incoherent predictions for that participant, divided by $n$, the total number of decision points of the participant. A predicted answer is *coherent* if it is the same as the participant's answer, otherwise it is *incoherent*. A *decision point* is a turn in a game where a participant has to give an answer. The parameter $p$ is a penalty coefficient, which is applied when a participant does *not* follow a certain ToM level, but *does* match its actions. We set it to 0.5. Predicted answers are generated as described at the end of Sect. 2.4. We deviate from [8], where models are fitted to full games instead of decision points. After all, participants can have multiple decision points in each game (one for each round).

In addition to ToM levels 0 through 5, we also fit a *random model*. We determine the best fitting random model by considering that each player guesses among the four options with a fixed but personal probability. The log-likelihood for the random model is

$$\sum_{a \in Ans} a \cdot \ln(\frac{a}{n})$$

where $n$ is the total number of decision points, and $Ans = (k_\neg, k_{88}, k_{8A}, k_{AA})$ is a list of numbers, where we define $k_\neg$ as the number of times the participant answered 'I do not know my cards', $k_{88}$ as the number of times the participant answered 'I know I have two Eights', et cetera.

Given these likelihoods, RFX-BMS estimates a vector $\alpha$, containing one element for each ToM level and an additional element for the random model.[4]

## 3   Results

In Sect. 3.1, we explore the use of RFX-BMS by combining it with the epistemically bounded models of Cedegao and colleagues [8], as outlined in Sect. 2.3. In Sect. 3.2, we use the TOMPAL models introduced in Sect. 2.4 as models in RFX-BMS (as described in Sect. 2.5), which we use to predict the frequencies of each ToM level in the data of [8].

### 3.1   Predicted Epistemic Levels of Participants

Before employing our novel models, we validate the use of RFX-BMS by using it to estimate the relative frequencies of epistemic levels for subjects in [8] by using as model a non-stochastic version of SUWEB, the best-fitting model in [8], which employs the bounded models described in Sect. 2.3. SUWEB models have an *update probability*, the probability with which a state is removed after an announcement, and a *noise* parameter, the probability of the model guessing 'I know' when it does *not* know. We set these to 1 and 0, respectively. When SUWEB considers no states to be possible, it answers 'I know' or 'I don't know'

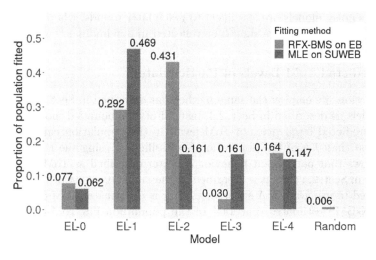

**Fig. 3.** In red, relative frequencies of each epistemic level and the random model as predicted by RFX-BMS, for [8]'s data, using bounded models. In blue, the original fit of [8]'s stochastic SUWEB models, which also are bounded models. (Color figure online)

---

[4] All code used for this article can be found at https://github.com/jdtoprug/EpistemicToMProject and doi: 10.5281/zenodo.8382660. Note that we implemented the model updates needed for Aces and Eights and related games, and not a general logical framework.

with equal probability. In these cases we have this non-stochastic SUWEB answer 'I don't know' instead. We combine this non-stochastic SUWEB with RFX-BMS as described in Sect. 2.5, in order to estimate the relative frequencies of each epistemic level, as well as the random model, across all 211 participants.

The predicted frequencies of epistemic levels in the population can be found in Fig. 3. Here, the blue bars are the original fit of [8], obtained by using Maximum Likelihood Estimation to estimate SUWEB's parameters and the epistemic level (EL) of each participant. The red bars are the predictions of RFX-BMS on non-stochastic SUWEB (EB), as explained in the previous paragraph. As a reminder, both red and blue bars use bounded models as explained in Sect. 2.3. For non-stochastic SUWEB, less than 1% of the population is classified as using the random model, which validates the epistemically bounded models presented in [8]. Over 40% of the population is classified as EL-2. This differs from the original SUWEB, which fits over 45% of participants to EL-1. We believe this is because many of the games that reportedly require levels 3 or 4 can be correctly solved by simply answering 'I don't know' in every round, which our non-stochastic EL-2 models consistently do, as opposed to the original SUWEB models, which sometimes answer 'I know' due to noise. Many participants that were fitted as EL-3 or EL-4 can be reclassified as EL-2 users who use this heuristic. For non-stochastic models, update probabilities are 1, which should make higher-level behavior less similar to lower-level behavior, as it causes models to say 'I don't know' less frequently. Zero noise may also decrease similarity between models, as noisy models are less likely to reach later rounds, where levels can be distinguished. These effects should be reflected in our findings.[5]

## 3.2 Predicted ToM Levels of Participants

In this section, we employ the same methods as described in Sect. 3.1, using our ToM models as described in Sect. 2.4, instead of [8]'s bounded models.

The predicted frequencies of ToM levels in the population can be found in Fig. 4. Less than 1% of the population is classified as using the random model, which shows that participant behavior is better described as ToM reasoning as described in Sect. 2.4 than it is described as guessing. Over 35% of the population is predicted to use ToM-2. A surprising result is the peak at ToM-5: it turns out that RFX-BMS estimates that 14% of the population fits ToM-5 better than any other ToM level. This is not dissimilar to [8], where 15% of participants is fitted to epistemic level 4 (the rightmost blue bar in Fig. 3). In our models, in order to solve *all* games, ToM-5 is needed, whereas in [8], non-stochastic EL-4 accomplishes the same.

When comparing the RFX-BMS results for the epistemically bounded and ToM models, we see that the estimated frequency of ToM-2 users is lower than that of EL-2 users. We believe this is because there are some games where non-stochastic EL-2 correctly answers 'I do not know my cards' due to becoming

---

[5] We cannot test these predictions as we do not have access to the computational power required to fit the SUWEB model of [8] in a reasonable amount of time.

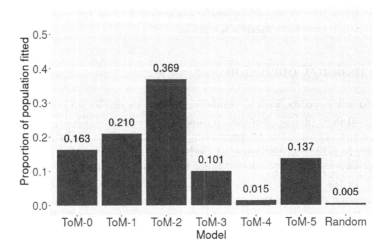

**Fig. 4.** Relative frequencies of each ToM level and the random model as predicted by RFX-BMS, for the data of [8], using ToM models.

'confused' and removing all non-peripheral nodes, whereas our ToM-2 models incorrectly answer 'I know my cards' due to mistakenly attributing ToM-1 to the other players (which are ToM-5). For one such example, see Appendix B.

To see how well, on average, our models' predictions correspond to participant behavior, the distribution of coherence across participants can be found in Fig. 5. A participant's coherence is the number of coherent predictions for that participant's best-fitting model, divided by that participant's total number of decision points. Coherence is at least .736 for over half of the participants, and only 15 participants have a coherence of 0.5 or lower. There are only six participants where the random model has the best coherence, which are indicated using an ×. Upon visual inspection of the data for the low-coherence outliers, it

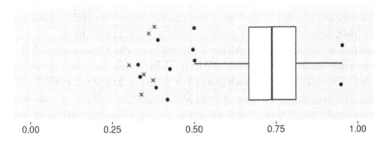

**Fig. 5.** Distribution of 1-ε for the best-fitting ToM levels for each of the 211 participants. Mean 0.723, median 0.737, IQR 0.143. Crosses indicate participants for whom the random model fits better than any of the ToM models. The vertical axis has no meaning and is used to separate data points for improved readability.

seems that these participants frequently answered 'I know my cards' when they could not, which our ToM models never do.

## 4   Discussion/Conclusion

Humans do not have the logical omniscience that modal logics based on Kripke models presuppose [21,39]. For one, human ToM is limited [23,27]. In this paper we propose a novel method of representing ToM limitations in Public Announcement Logic, building on the work of Cedegao et al. [8] (see also [22] and [11]). We use Random-Effects Bayesian Model Selection to predict the frequencies of ToM levels in the data of [8], and find some striking differences and similarities when comparing the estimates of ToM and epistemically bounded models.

We predict the majority of the participants of Cedegao and colleagues [8] to be using ToM-2, possibly bolstered by the heuristic of answering 'I don't know' in cases where a random answer would be given in the SUWEB model of [8]. For the latter, the majority of participants is fitted as Cedegao et al.'s epistemic level 1 (EL-1). We believe this difference is due to SUWEB's stochasticity, as well as EL-2 and higher overestimating human (recursive) reasoning capabilities. Our results are a refinement that show that participants are better described as ToM-2 than ToM-1, where the former lies between non-stochastic EL-1 and EL-2 in terms of game-solving capabilities. Our novel method also predicts a portion of participants to use ToM-5. However, since participants can solve many higher-level puzzles by always answering 'I don't know', it is difficult to distinguish higher-order reasoning from heuristics, so it is important to emphasize that the participants themselves may not necessarily be using fifth-order reasoning. We recommend employing games where to be correct, one must eventually answer 'I know' as diagnostic cases in future research.

A drawback of our approach is that we do not consider deviations from our ToM models' predictions, even though some participants exhibit clear guessing strategies where they answer 'I know my cards' when they cannot know. Also, our models do not consider the possibility that agents may attribute different levels of ToM reasoning to other players. For example, a ToM-2 model attributes ToM-1 to every other agent, and does not consider the possibility that one agent is using ToM-0, whereas another agent is using ToM-1. Furthermore, we assume that participants use a single ToM level throughout the experiment, but it could be possible that some participants switch ToM levels between games or even rounds. Lastly, recall that our models answer 'I do not know my cards' when there are no outgoing edges. When this answer is changed to a different answer, or any random distribution over the four answers, we find that mean coherence never drops under 0.72. However, we assume that all participants use the same strategy in such cases, whereas a richer model could try to find the best-fitting answering behavior for each player. In future work it may be possible to incorporate all these behaviors in our models, though even without covering these cases our models have a mean 0.723 coherence - a decent fit, and an indication that participant behavior can, at least partially, be described using our ToM models.

In Sect. 2.5, we calculate the log-likelihood of a ToM level fitting a participant by introducing a penalty for deviating from our models, the value of which strongly affects the relative fit of the random model compared to ToM models. Random-Effects Bayesian Model Selection must assign each participant to one of our defined models. Though we included ToM and random models, there may be other models that fit even better. For example, participants may be using a representation similar to the number triangles in [15], they may be generalizing such as the participants in [18], or they may be using other strategies. More research and data is needed to find all relevant behavioral features. Eye-tracking data could be used to distinguish between strategies, allowing for more accurate logically inspired models [26,34,35]. These models need not be based on formal logics: we also encourage cognitive scientists to model higher-order ToM in Aces and Eights. Nonetheless, we demonstrate that a large part of participant behavior can be attributed to ToM limitations as represented in our models.

**Acknowledgements.** This research was funded by the project 'Hybrid Intelligence: Augmenting Human Intellect', a 10-year Gravitation programme funded by the Dutch Ministry of Education, Culture and Science through the Netherlands Organisation for Scientific Research, grant number 024.004.022. Lastly, we would like to thank our four anonymous reviewers and prof. dr. Hans van Ditmarsch for providing us with helpful comments, suggestions, and discussion.

## Appendix A

This appendix describes how to extend our work beyond Aces and Eights.

In [22], concatenation of sequences is defined: $e \circ e' = (i_1, \ldots, i_m, j_1, \ldots, j_k)$ for $e = (i_1, \ldots, i_m)$, $e' = (j_1, \ldots, j_k)$. The empty sequence is $\epsilon$, and $e \circ \epsilon = \epsilon \circ e = e$.

The epistemic depth $\delta(F)$ of a formula $F$ is inductively defined as follows:

D0: $\delta(p) = \{\epsilon\}$ for any $p \in P$;
D1: $\delta(\neg F) = \delta(F)$;
D2: $\delta(F \rightarrow G) = \delta(F) \cup \delta(G)$;
D3: $\delta(\wedge \Phi) = \delta(\vee \Phi) = \cup_{F \in \Phi} \delta(F)$;
D4: $\delta(K_i(F)) = \{(i) \circ e : e \in \delta(F)\}$.
D5: $\delta([F]G) = \{f \circ e : e \in \delta(F), f \in \delta(G)\}$

We added D5, which is not present in [22]. Moving to novel work, we define the ToM structure $\mathcal{T}_{(p,l)}$, with $p \in A$ and $l \in \mathbb{N}$ inductively as follows:

> **Base Case:** $e \in \mathcal{T}_{(p,l)}$ for every $e = (i_1, \ldots, i_m)$ where $0 \leq m \leq l$, and for every $i_j \in e$ we have that $i_j \in A$ and [if $0 < j < m$, then $i_j \neq i_{j+1}$]. If $m \leq 0$ then $e = \epsilon$.
> **Inductive Step 1:** If $e \in \mathcal{T}_{(p,l)}$ and $l \geq 0$, then $(p) \circ e \in \mathcal{T}_{(p,l)}$
> **Inductive Step 2:** If, for any $e_1, i, e_2$; $e_1 \circ ((i) \circ e_2) \in \mathcal{T}_{(p,l)}$, then $(e_1 \circ (i)) \circ ((i) \circ e_2) \in \mathcal{T}_{(p,l)}$

Our base case corresponds to our requirement that the number of 'perspective switches' is limited by an agent's ToM order. Inductive steps 1 and 2 correspond to not switching perspectives, not requiring additional ToM.

For zero or more repetitions of $i$ we write $i^*$. As an example, consider $A = \{0,1\}$. Then, $\mathcal{T}_{(0,2)} = \{\epsilon, (0^*), (1^*), (0^*, 1^*), (1^*, 0^*), (0^*, 1^*, 0^*)\}$.

We then modify our semantic definition of $[\varphi]\psi$ in Definition 3:

$$M, (s, (i, l)) \models [\varphi]\psi \quad \Leftrightarrow \quad M, (s, (i, l)) \models \varphi \text{ implies } M|\varphi, (s, (i, l)) \models \psi$$

where we define the model restriction $M|\varphi = (S, R, V, T')$ with $(i, l) \in T'(s)$ iff $(i, l) \in T(s)$ and $[M, (s, (i, l)) \models \varphi$ or $[\delta(\varphi) \not\subseteq \mathcal{T}_{(i,l)}]]$.

Note that $\delta(\varphi) \not\subseteq \mathcal{T}_{(i,0)}$ is equivalent to "$\varphi$ contains an operator $K_j$ with $i \neq j$", as $\mathcal{T}_{(i,0)} = \{\epsilon, (i^*)\}$. With this substitution, our proofs for Theorems 1–3 hold, and our models can be used with any announcements.

# Appendix B

There are two games where non-stochastic EL-2 answers correctly whereas our ToM-2 models answer incorrectly.[6] In both of these, the participant is player 0. The distribution of cards in these games is $AA8A88$ and $8A8AAA$. For the

Table 1. Tuples at each relevant state during a series of announcements.

| AA8888 | **AA8A88** | AAAA88 | 8AAA88 | 88AA88 | 8A8A88 | next |
|---|---|---|---|---|---|---|
| 0: 0,1,2,3,4,5 | 0: 0,1,2,3,4,5 | 0: 0,1,2,3,4,5 | 0: 0,1,2,3,4,5 | 0: 0,1,2,3,4,5 | 0: 0,1,2,3,4,5 | |
| 1: 0,1,2,3,4,5 | 1: 0,1,2,3,4,5 | 1: 0,1,2,3,4,5 | 1: 0,1,2,3,4,5 | 1: 0,1,2,3,4,5 | 1: 0,1,2,3,4,5 | 0: $k\neg$ |
| 2: 0,1,2,3,4,5 | 2: 0,1,2,3,4,5 | 2: 0,1,2,3,4,5 | 2: 0,1,2,3,4,5 | 2: 0,1,2,3,4,5 | 2: 0,1,2,3,4,5 | |
| 0: | 0: 0,1,2,3,4,5 | 0: 0,1,2,3,4,5 | 0: 0,1,2,3,4,5 | 0: 0,1,2,3,4,5 | 0: 0,1,2,3,4,5 | |
| 1: 0 | 1: 0,1,2,3,4,5 | 1: 0,1,2,3,4,5 | 1: 0,1,2,3,4,5 | 1: 0,1,2,3,4,5 | 1: 0,1,2,3,4,5 | 1: $k\neg$ |
| 2: 0 | 2: 0,1,2,3,4,5 | 2: 0,1,2,3,4,5 | 2: 0,1,2,3,4,5 | 2: 0,1,2,3,4,5 | 2: 0,1,2,3,4,5 | |
| 0: | 0: 0,1,2,3,4,5 | 0: 0,1,2,3,4,5 | 0: 0,1,2,3,4,5 | 0: 0 | 0: 0,1,2,3,4,5 | |
| 1: 0 | 1: 0,1,2,3,4,5 | 1: 0,1,2,3,4,5 | 1: 0,1,2,3,4,5 | 1: | 1: 0,1,2,3,4,5 | 2: $k\neg$ |
| 2: 0 | 2: 0,1,2,3,4,5 | 2: 0,1,2,3,4,5 | 2: 0,1,2,3,4,5 | 2: 0 | 2: 0,1,2,3,4,5 | |
| 0: | 0: 0,1,2,3,4,5 | 0: 0 | 0: 0,1,2,3,4,5 | 0: 0 | 0: 0,1,2,3,4,5 | |
| 1: 0 | 1: 0,1,2,3,4,5 | 1: 0 | 1: 0,1,2,3,4,5 | 1: | 1: 0,1,2,3,4,5 | 0: $k\neg$ |
| 2: 0 | 2: 0,1,2,3,4,5 | 2: | 2: 0,1,2,3,4,5 | 2: 0 | 2: 0,1,2,3,4,5 | |
| 0: | 0: 0,1,2,3,4,5 | 0: 0 | 0: 0 | 0: 0 | 0: 0,1,2,3,4,5 | |
| 1: 0 | 1: 0,1,2,3,4,5 | 1: 0 | 1: 0,1 | 1: | 1: 0,1,2,3,4,5 | 1: $k$ |
| 2: 0 | 2: 0,1,2,3,4,5 | 2: | 2: 0,1 | 2: 0 | 2: 0,1,2,3,4,5 | |
| 0: | 0: 0,  2,3,4,5 | 0: 0 | 0: 0 | 0: 0 | 0: 0,  3,4,5 | |
| 1: | 1:  1,2,3,4,5 | 1: | 1: | 1: | 1:  2,3,4,5 | 2: $k\neg$ |
| 2: 0 | 2: 0,  2,3,4,5 | 2: | 2: 0 | 2: 0 | 2: 0,  3,4,5 | |
| 0: | 0: 0,  2,  4,5 | 0: 0 | 0: 0 | 0: 0 | 0: 0,  3,4,5 | |
| 1: | 1:  1,2,  4,5 | 1: | 1: | 1: | 1:  2,3,4,5 | |
| 2: 0 | 2: 0,   3,4,5 | 2: | 2: 0 | 2: 0 | 2: 0,   3,4,5 | |

---

[6] Because knowledge can be false, using 'knowledge' and $K$ may not be entirely accurate. We use it because the model for Aces and Eights is $S5$, but for future work we recommend using 'beliefs' and $B$.

former, we show the removal of tuples after each announcement in Table 1, where each column is a relevant state, and each row corresponds to an announcement. Column ordering corresponds to the order of states in Fig. 1. The rightmost column shows the next announcement, where the index denotes the player, $k$ is 'I know my cards', and $k\neg$ is 'I do not know my cards'. Tuples that will be removed after the next announcement are red. After six announcements, player 0 at ToM-2 will incorrectly answer 'I know my cards', whereas at ToM-5 she will answer 'I do not know my cards', which is the correct answer. When working through the example, it is recommended to use Fig. 1 as a companion.

# References

1. Arslan, B., Verbrugge, R., Taatgen, N., Hollebrandse, B.: Accelerating the development of second-order false belief reasoning: a training study with different feedback methods. Child Dev. **91**(1), 249–270 (2020). https://doi.org/10.1111/cdev.13186
2. Arthaud, F., Rinard, M.: Depth-bounded epistemic logic. In: Verbrugge, L.C. (ed.) Proceedings of the 19th Conference on Theoretical Aspects of Rationality and Knowledge (TARK 23), pp. 46–65 (2023). https://doi.org/10.4204/EPTCS.379.7
3. Baltag, A., Moss, L.S., Solecki, S.: The logic of public announcements, common knowledge, and private suspicions. In: Gilboa, I. (ed.) Proceedings of the 7th Conference on Theoretical Aspects of Rationality and Knowledge (TARK 98), pp. 43–46 (1998)
4. Baltag, A., Moss, L.S., Solecki, S.: The logic of public announcements, common knowledge, and private suspicions. In: Arló-Costa, H., Hendricks, V.F., van Benthem, J. (eds.) Readings in Formal Epistemology. SGTP, vol. 1, pp. 773–812. Springer, Cham (2016). https://doi.org/10.1007/978-3-319-20451-2_38
5. Barrett, J.L., Richert, R.A., Driesenga, A.: God's beliefs versus mother's: The development of nonhuman agent concepts. Child Dev. **72**(1), 50–65 (2001). https://doi.org/10.1111/1467-8624.00265
6. Blackburn, P., De Rijke, M., Venema, Y.: Modal logic (Cambridge tracts in theoretical computer science no. 53). Cambridge University Press (2001)
7. Camerer, C.F., Ho, T.H., Chong, J.K.: A cognitive hierarchy model of games. Q. J. Econ. **119**(3), 861–898 (2004). https://doi.org/10.1162/0033553041502225
8. Cedegao, Z., Ham, H., Holliday, W.H.: Does Amy know Ben knows you know your cards? A computational model of higher-order epistemic reasoning. In: Proceedings of the 43th Annual Meeting of the Cognitive Science Society, pp. 2588–2594 (2021)
9. De Weerd, H., Diepgrond, D., Verbrugge, R.: Estimating the use of higher-order theory of mind using computational agents. BE J. Theor. Econ. **18**(2) (2018). https://doi.org/10.1515/bejte-2016-0184
10. De Weerd, H., Verbrugge, L.C., Verheij, B.: How much does it help to know what she knows you know? An agent-based simulation study. Artif. Intell. **199–200**, 67–92 (2013). https://doi.org/10.1016/j.artint.2013.05.004
11. Dégremont, C., Kurzen, L., Szymanik, J.: Exploring the tractability border in epistemic tasks. Synthese **191**(3), 371–408 (2014). https://doi.org/10.1007/s11229-012-0215-7
12. Devaine, M., Hollard, G., Daunizeau, J.: The social Bayesian brain: Does mentalizing make a difference when we learn? PLoS Comput. Biol. **10**(12), e1003992 (2014). https://doi.org/10.1371/journal.pcbi.1003992

13. Etel, E., Slaughter, V.: Theory of mind and peer cooperation in two play contexts. J. Appl. Dev. Psychol. **60**, 87–95 (2019). https://doi.org/10.1016/j.appdev.2018. 11.004

14. Fagin, R., Halpern, J.Y., Moses, Y., Vardi, M.: Reasoning About Knowledge. MIT Press, Cambridge (1995)

15. Gierasimczuk, N., Szymanik, J.: A note on a generalization of the muddy children puzzle. In: Proceedings of the 13th Conference on Theoretical Aspects of Rationality and Knowledge, pp. 257–264 (2011). https://doi.org/10.1145/2000378.2000409

16. Goodie, A.S., Doshi, P., Young, D.L.: Levels of theory-of-mind reasoning in competitive games. J. Behav. Decis. Mak. **25**(1), 95–108 (2012). https://doi.org/10. 1002/bdm.717

17. Hall-Partee, B.: Semantics-mathematics or psychology? In: Bäuerle, R., Egli, U., Von Stechow, A. (eds.) Semantics from Different Points of View, SSLC, vol. 6, pp. 1–14. Springer, Berlin, Heidelberg (1979). https://doi.org/10.1007/978-3-642-67458-7_1

18. Hayashi, H.: Possibility of solving complex problems by recursive thinking. Jpn. J. Psychol. **73**(2), 179–185 (2002). https://doi.org/10.4992/jjpsy.73.179

19. Hintikka, J.: Knowledge and Belief: An Introduction to the Logic of the Two Notions. Cornell University Press, Ithaca, NY, USA (1962)

20. Jonker, C.M., Treur, J.: Modelling the dynamics of reasoning processes: Reasoning by assumption. Cogn. Syst. Res. **4**(2), 119–136 (2003). https://doi.org/10.1016/S1389-0417(02)00102-X

21. Kahneman, D., Slovic, P., Tversky, A.: Judgment under Uncertainty: Heuristics and Biases. Cambridge University Press, Cambridge (1982)

22. Kaneko, M., Suzuki, N.Y.: Epistemic logic of shallow depths and game theoretical applications. In: Advances In Modal Logic, vol. 3, pp. 279–298. World Scientific (2002). https://doi.org/10.1142/9789812776471_0015

23. Keysar, B., Lin, S., Barr, D.J.: Limits on theory of mind use in adults. Cognition **89**(1), 25–41 (2003). https://doi.org/10.1016/S0010-0277(03)00064-7

24. Kline, J.J.: Evaluations of epistemic components for resolving the muddy children puzzle. Econ. Theor. **53**(1), 61–83 (2013). https://doi.org/10.1007/s00199-012-0735-x

25. McCarthy, J.: Formalization of two puzzles involving knowledge. Formalizing Common Sense: Papers by John McCarthy, pp. 158–166 (1990)

26. Meijering, B., van Rijn, H., Taatgen, N.A., Verbrugge, R.: What eye movements can tell about theory of mind in a strategic game. PLoS ONE **7**(9), 1–8 (2012). https://doi.org/10.1371/journal.pone.0045961

27. Nagel, R.: Unraveling in guessing games: An experimental study. Am. Econ. Rev. **85**(5), 1313–1326 (1995)

28. Paal, T., Bereczkei, T.: Adult theory of mind, cooperation, Machiavellianism: The effect of mindreading on social relations. Pers. Individ. Differ. **43**(3), 541–551 (2007). https://doi.org/10.1016/j.paid.2006.12.021

29. Plaza, J.: Logics of public announcements. In: Emrich, M., Pfeifer, M., Hadzikadic, M., Ras, Z. (eds.) Proceedings of the 4th International Symposium on Methodologies for Intelligent Systems: Poster Session Program, pp. 201–216. Oak Ridge National Laboratory (1989)

30. Premack, D., Woodruff, G.: Does the chimpanzee have a theory of mind? Behav. Brain Sci. **1**(4), 515–526 (1978). https://doi.org/10.1017/S0140525X00076512

31. Solaki, A.: The effort of reasoning: modelling the inference steps of boundedly rational agents. J. Log. Lang. Inform. **31**(4), 529–553 (2022). https://doi.org/10. 1007/s10849-022-09367-w

32. Stahl, D.O., II., Wilson, P.W.: Experimental evidence on players' models of other players. J. Econ. Behav. Organ. **25**(3), 309–327 (1994). https://doi.org/10.1016/0167-2681(94)90103-1

33. Stephan, K.E., Penny, W.D., Daunizeau, J., Moran, R.J., Friston, K.J.: Bayesian model selection for group studies. Neuroimage **46**(4), 1004–1017 (2009). https://doi.org/10.1016/j.neuroimage.2009.03.025

34. Top, J.D., Verbrugge, R., Ghosh, S.: An automated method for building cognitive models for turn-based games from a strategy logic. Games **9**(3), 44 (2018). https://doi.org/10.3390/g9030044

35. Top, J.D., Verbrugge, R., Ghosh, S.: Automatically translating logical strategy formulas into cognitive models. In: 16th International Conference on Cognitive Modelling, pp. 182–187 (2018)

36. Van Ditmarsch, H.: Dynamics of lying. Synthese **191**(5), 745–777 (2014)

37. Van Ditmarsch, H., van der Hoek, W., Kooi, B.: Dynamic Epistemic Logic, Synthese Library, vol. 337. Springer Science & Business Media, Dordrecht, Netherlands (2007). https://doi.org/10.1007/978-1-4020-5839-4

38. Veltman, K., de Weerd, H., Verbrugge, R.: Training the use of theory of mind using artificial agents. J. Multimodal User Interfaces **13**(1), 3–18 (2019). https://doi.org/10.1007/s12193-018-0287-x

39. Verbrugge, R.: Logic and social cognition: The facts matter, and so do computational models. J. Philos. Log. **38**(6), 649–680 (2009). https://doi.org/10.1007/s10992-009-9115-9

40. Verbrugge, R., Meijering, B., Wierda, S., Van Rijn, H., Taatgen, N.: Stepwise training supports strategic second-order theory of mind in turn-taking games. Judgm. Decis. Mak. **13**(1), 79–98 (2018). https://doi.org/10.1017/S1930297500008846

41. Wimmer, H., Perner, J.: Beliefs about beliefs: Representation and constraining function of wrong beliefs in young children's understanding of deception. Cognition **13**(1), 103–128 (1983). https://doi.org/10.1016/0010-0277(83)90004-5

# Learning by Intervention in Simple Causal Domains

Katrine Bjørn Pedersen Thoft and Nina Gierasimczuk[(✉)] [iD]

Department of Applied Mathematics and Computer Science,
Technical University of Denmark, Kongens Lyngby, Denmark
{kabpt,nigi}@dtu.dk

**Abstract.** We propose a framework for learning dependencies between variables in an environment with causal relations. We assume that the environment is fully observable and that the underlying causal structure is of a simple nature. We adapt the frameworks of the (epistemic) causal models from [4,17], and propose a model inspired by action learning [6,7]. We present two learning methods, using formal and algorithmic approaches. Our learning agents infer dependencies (atomic formulas of Dependence Logic) from observations of interventions on valuations (propositional states), and by doing so efficiently, they obtain insights into how to manipulate their surroundings to achieve goals.

**Keywords:** causality · causal models · dependence models · dependence logic · learning by intervention · finite identifiability · artificial intelligence

## 1 Introduction

In this paper, we set the stage for a model of learning dependencies of cause-effect relationships. Our perspective is formal and algorithmic, and as such it contributes to the architecture of artificial agents. Causal inference is also of paramount importance in epistemology, in philosophy of science (as discovering cause-effect relations is one of the fundamental tasks of empirical sciences), and in cognitive science. The concept of causality appears in cognition quite early in human development—children as young as the age of six months are able to identify some categories of cause-effect relations [16].

Studying causation formally is very challenging, as it can be easily confused with correlation. Pearl [18] proposes three practical levels of analysis, the so-called 'Ladder of Causality': prediction, manipulation, and counterfactuals. On the first rung, prediction, agents can only observe the environment and make predictions of outcomes, while the second rung agents can make predictions of how their actions affect the environment. In the third and last rung of the ladder, counterfactuals, agents can imagine hypothetical scenarios in the environment and predict outcomes. In this paper we focus on the second level, the level of *manipulation* (or *intervention*). We consider an agent executing actions

N. Gierasimczuk and F. R. Velázquez-Quesada (Eds.): DaLí 2023, LNCS 14401, pp. 104–118, 2024.
https://doi.org/10.1007/978-3-031-51777-8_7

in an environment that functions according to an unknown causal structure. The agent's goal is to learn (infer) the dependencies between variables in the environment. While our causal structure says explicitly how the values of certain variables influence the values of other variables, dependency structure is less specific—it only points to the existence of a relationship between variables, without specifying its nature exactly. Our methodology borrows from modal logic-based interpretation of learning [9], and is closely related to the learning of Dynamic Epistemic Logic action models [6,7], where agents learn to predict the effects of actions, but they do not address the dependence between variables explicitly. In the present paper we draw inspiration from the recent work on (epistemic) causal models [4] and Dependence Logic [20]. In fact, our agent learns atomic Dependence Logic formulas describing sets of valuations compliant to the causal structural functions of the domain.

## 2   Modelling Simple Causality

According to Von Wright, two propositions are causally connected if *we can influence one by manipulating the other*. He calls this type of causal connection *manipulative* causation, as it points to an essential connection between causation and action [21]. Recent work in developmental cognitive psychology reveals that children indeed use information from their interventions to correctly disambiguate the structure of a causal chain [15]. We will adopt these intuitions in our model and distinguish a special kind of *manipulable* variables, the value of which can be directly changed by the agent.

*Example 1.* Consider a simple train-track control set-up, see Fig. 1. The agent finds herself in an environment with two levers, a red one ($r$) and a blue one ($b$), and two train tracks. The underlying causal relationships are as follows: pulling down both levers causes the tracks to merge ($m$), pulling down the red lever causes the traffic to stop ($t$).

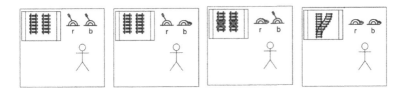

**Fig. 1.** The train-track control domain

The causal models of Pearl [17] distinguish between the causally independent *exogenous* variables (like $r$ and $b$ in our example), and the causally dependent *endogenous* variables (like $t$ and $m$). In this paper we will make several simplifying assumptions. We will take the set of exogenous variables to be equal to the manipulable variables, i.e., the agent can manipulate all and only exogenous variables. Moreover, our variables are binary—they only take values from the set $\{0,1\}$. Finally, we only consider *simple causality* of chains of length 2, i.e., endogenous variables cannot affect other endogenous variables. We will discuss the consequences of lifting these assumptions in the concluding remarks, in Sect. 6.

Our starting point is the notion of a *causal frame*, which codes the structure of the environment: the variables and how they truth-functionally affect each other.

**Definition 1 (Simple Causal Frame).** *Let $U = \{u_0, \ldots, u_{n-1}\}$ and $V = \{v_0, \ldots, v_{k-1}\}$ be (disjoint) sets of exogenous and endogenous variables, respectively. A simple causal frame over $U$ and $V$ is a tuple $\mathbb{C} = (U, V, \mathcal{F})$, where $\mathcal{F}$ assigns a map $f_{v_j} : \{0,1\}^n \to \{0,1\}$ to each endogenous variable $v_j \in V$, i.e., for each valuation of all exogenous variables in $U$, $f_{v_j}$ determines the value of the endogenous variable $v_j$. Given (disjoint) sets of exogenous and endogenous variables $U$ and $V$, we define $\mathrm{CF}(U,V)$ as the set of all simple causal frames over $U$ and $V$.*

Direct causal influence of a variable $u$ over a variable $v$ requires that there exists a valuation within which just the change of the value of $u$ triggers a change of the value of $v$.

**Definition 2 (Causal Influence).** *Let $\mathbb{C} = (U, V, \mathcal{F})$ be a simple causal frame. We say that an endogenous variable $v_j \in V$ is directly causally influenced by an exogenous variable $u_i \in U$ if and only if there is a valuation $g : U \to \{0,1\}$, such that:*

$$f_{v_j}(g(u_0), \ldots, g(u_i), \ldots, g(u_{n-1})) \neq f_{v_j}(g(u_0), \ldots, 1 - g(u_i), \ldots g(u_{n-1})).$$

A particular 'instance' of a causal frame, i.e., a frame with a distinguished valuation, will be called a *causal model*.

**Definition 3 (Simple Causal Model).** *A simple causal model is a tuple $\mathbb{C} = (U, V, \mathcal{F}, a)$, where: $(U, V, \mathcal{F})$ is a causal frame, $U = \{u_0, \ldots, u_{n-1}\}$ and $a : U \cup V \to \{0,1\}$ is a valuation that complies with $\mathcal{F}$, i.e., for all $v_j \in V$, $a(v_j) = f_{v_j}(a(u_0), \ldots, a(u_{n-1}))$.*

Causal models allow for modelling interventions: by manipulating the values of variables the agent 'jumps' between causal models, as the distinguished, actual valuation changes. As mentioned before, in our framework (all and only) exogenous variables of the model can be manipulated. This leads to the following notion of intervention, adapted from [4].

**Definition 4 (Intervention).** *Given $U = \{u_0, \ldots, u_{n-1}\}$ and $V$, a simple causal model $\mathbb{C} = (U, V, \mathcal{F}, a)$, with $u_i \in U$, and $x \in \{0,1\}$, the intervention is $\mathbb{C}_{u_i := x} = (U, V, \mathcal{F}, a_{u_i := x})$, where*

$a_{u_i:=x} : U \cup V \to \{0,1\}$ *is such that:*

$$a_{u_i:=x}(y) = \begin{cases} x & \text{if } y = u_i; \\ a(y) & \text{if } y \in U \setminus \{u_i\}; \\ f_v(a(u_0), \ldots, a(u_i) := x, \ldots, a(u_{n-1})) & \text{if } y = v \in V. \end{cases}$$

Note that an intervention need not change the existing valuation, i.e., when the newly assigned value is the same as the old one. Such interventions will be called *trivial.*

Intuitively, an intervention will only change the value of the affected exogenous variable itself and, in accordance with $\mathcal{F}$, all endogenous variables that are directly causally influenced by this exogenous variable.

# 3  Modelling Simple Dependence

One symptom of causation is dependence between variables. When talking about dependence, several existing paradigms should be mentioned. The first one, Independence Friendly Logic [14] is an extension of First-Order Logic, where independence is treated on the quantifier level. Jouko Väänänen's Dependence Logic (DL, [20]) and its propositional version [22] treat dependencies on the atomic level of formulas. Generalizing Tarski's semantics, it's interpreted on so-called 'teams', i.e., sets of valuations. For completeness, it is also important to mention the recently proposed logic of functional dependence [1], which addresses two basic kinds of dependence: the global and the local one. In this paper, we are interested in studying dependence between boolean variables. We adopt a perspective close to that of DL, where dependence is expressed using the dependence atom $=(x_1, \ldots, x_n, y)$ read as: *the value of $y$ depends on the values of $x_1, \ldots, x_n$.* Formally, the meaning of this expression is defined within team semantics in the following way: we say that a set of valuations $\mathcal{X}$ is of type $=(x_0, \ldots, x_k, v)$ iff for all $a, b \in \mathcal{X}$ we have that if $a(x_0) = b(x_0), \ldots, a(x_k) = b(x_k)$, then $a(v) = b(v)$.

**Definition 5 (Simple Dependence Model).** *Let $U$ and $V$ be disjoint sets of exogenous and endogenous variables, respectively. A* simple dependence model *is a triple $\mathbf{D} = (U, V, F)$, where $F \subseteq \mathcal{P}(U \cup V)$ is the smallest set such that for each $v \in V$ there is a unique $U' \subseteq U$ with $U' \cup \{v\} \in F$ (we will refer to each such element with $F_v$).*

*Given (disjoint) sets of exogenous and endogenous variables $U$ and $V$, we define $\mathrm{DM}(U, V)$ as the set of all simple dependence models over $U$ and $V$.*

A dependence model can be seen as a coarser representation of causality. It does not specify exactly *how* a given endogenous variable is influenced by exogenous variables. Instead, it just lists the relevant exogenous variables that determine it. There is hence many-to-one correspondence between causal frames and dependence models, given by the following definition.

**Definition 6.** *For any simple causal frame* $\mathbb{C} = (U, V, \mathcal{F})$, *the corresponding simple dependence model is* $\mathbf{D}^{\mathbb{C}} = (U, V, F)$, *where* $F$ *consists of sets* $F_{v_j}$, *one for each endogenous variable* $v_j \in V$, *that contains the variable* $v_j$ *itself, together with all and only exogenous variables that directly causally influence* $v_j$ *in* $\mathbb{C}$.

**Proposition 1.** *Let* $\mathbb{C} = (U, V, \mathcal{F})$ *and* $\mathbf{D}^{\mathbb{C}} = (U, V, F)$, *with* $F_v \in F$. *If we have that* $F_v = \{x_0, \ldots, x_k, v\}$, *then the set of valuations that comply with* $\mathcal{F}$ *is of type* $=(x_0, \ldots, x_k, v)$.

PROOF. Take $\mathbb{C} = (U, V, \mathcal{F})$. Assume that in $\mathbf{D}^{\mathbb{C}} = (U, V, F)$, $F_v = \{x_0, \ldots, x_k, v\}$.

We need to show that the set of valuations that comply with $\mathcal{F}$ is of type $=(x_0, \ldots, x_k, v)$, i.e., that for any two valuations $a$ and $b$ if $a(x_0) = b(x_0)$, ..., $a(x_k) = b(x_k)$, then $a(v) = b(v)$. Let us take two arbitrary valuations $a$ and $b$ that comply with $\mathcal{F}$, and for contradiction assume that $a(x_0) = b(x_0), \ldots, a(x_k) = b(x_k)$, but $a(v) \neq b(v)$. By the assumption of compliance with $\mathcal{F}$, we have that $a(v) = f_v(a(u_0), \ldots, a(u_{n-1}))$, and $b(v) = f_v(b(u_0), \ldots, b(u_{n-1}))$, so there must be a subset $Y \subseteq U \setminus \{x_0, \ldots, x_k\}$, such that for all $u \in Y$, $a(u) \neq b(u)$. We will argue that then there exists a single variable $u' \in Y$ that directly causally influences $v$ (but $u' \notin \{x_0, \ldots x_k\}$, which will give contradiction). To this end we need to construct a valuation $g : U \to \{0, 1\}$, such that

$$f_v(g(u_0), \ldots, g(u'), \ldots, g(u_{n-1})) \neq f_v(g(u_0), \ldots, 1 - g(u'), \ldots, g(u_{n-1})).$$

Let $Y = y_0, \ldots, y_\ell$. We construct a sequence of valuations $g_0, \ldots, g_\ell$ inductively in the following way:

$$g_0(x) := a(x)$$

$$g_{i+1}(x) := \begin{cases} 1 - g_i(x) & \text{if } x = y_i; \\ g_i(x) & \text{otherwise.} \end{cases}$$

The valuation we seek is the $g_i$ of the smallest $i$ such that:

$$f_v(g_i(u_0), \ldots, g_i(y_{i+1}), \ldots, g_i(u_{n-1})) \neq f_v(g_i(u_0), \ldots, 1 - g_i(y_{i+1}), \ldots, g_i(u_{n-1})).$$

Such an $i$ exists, since $g_\ell = b$.    $\square$

We are now well-equipped to introduce our learning framework. Our agents reside in a *causal frame*. By manipulating the values of variables in the frame, they 'jump' from one *causal model* to another. By observing the changes (pairs of such models), they learn which variables depend on each other. Storing all causal relations explicitly would require a lot of memory, so we only require they identify the (coarser) *dependence model* corresponding to the causal frame they are in. We argue that this concise partial knowledge is already useful enough to interact with the environment in an informed way. Knowing the dependence model corresponding to the causal frame, by Proposition 1, amounts to knowing the *type of the team* (expressed as an atomic formula of dependence logic) complying to the rules of the causal frame being learned.

## 4   Learning Dependencies in Causal Frames

We will now move on to learning dependence models that correspond (in the strict sense defined above) to the causal frames the agent intervenes with.

Let $\mathbb{C} = (U, V, \mathcal{F})$ be a simple causal frame, with $U = \{u_0, \ldots, u_{n-1}\}$ and $V = \{v_0, \ldots, v_{k-1}\}$, i.e., there are $n$ exogenous (manipulable) variables, and $k$ endogenous variables. A simple causal model $\mathbf{C} = (U, V, \mathcal{F}, a)$ can be understood as a *state* of the simple causal frame $\mathbb{C}$ given by the valuation $a$:

$$s_a = (a(u_0), \ldots, a(u_{n-1}), a(v_0), \ldots, a(v_{k-1})),$$

i.e., a binary sequence of length $n + k$ of values of all variables in $U \cup V$ under the valuation $a$. The enumeration order of variables in observations is the same and known to the agent throughout the learning process. Given the causal frame $\mathbb{C} = (U, V, \mathcal{F})$, we define the set of all possible states of $\mathbb{C}$ as $S_{\mathcal{F}} = \{s_b \mid b \text{ complies with } \mathcal{F}\}$.

Our learning function will output a dependence model given a finite sequence of *observations of interventions*, i.e., pairs $(s_b, s_c) \in S_{\mathcal{F}} \times S_{\mathcal{F}}$:

$$L : (S_{\mathcal{F}} \times S_{\mathcal{F}})^* \to \mathrm{DM}(U, V) \cup \{\uparrow\},$$

where $\uparrow$ stands for 'undecided'. Each intervention pair in an observation shows the state of the domain before and after an intervention, as in the learning model proposed in [6]. We thus need not impose a specific ordering on the observations for the learner to identify the dependence model.

The long-term behaviour of the learner will be defined with respect to a *stream* $\varepsilon \in (S_{\mathcal{F}} \times S_{\mathcal{F}})^{\infty}$, which is an infinite sequence of observations of interventions (repetitions are allowed). For $n \in \mathbb{N}$ and a stream $\varepsilon$ we use the following notation: $\varepsilon[n]$ is the initial segment of $\varepsilon$ of length $n + 1$; $\varepsilon_n$ is the $n$-th element of $\varepsilon$.

**Definition 7. (Stream and sequence for $\mathbb{C}$).** *Let $\mathbb{C} = (U, V, \mathcal{F})$ and $\beta$ is a finite sequence or a stream of observations of interventions, we will say that $\beta$ is for $\mathbb{C}$ if:*

1. *for all $n \in \mathbb{N}$, if $\beta_n = (s_b, s_c)$ then there is $u_i \in U$ and $x \in \{0, 1\}$ such that $c = b_{u_i := x}$;*
2. *for all $u_i \in U$ and for all $x \in \{0, 1\}$ there is an $n \in \mathbb{N}$ such that $\beta_n = (s_b, s_c)$ with $c = b_{u_i := x}$.*

In other words, a stream for a frame lists all possible interventions and their effects, and nothing more. As such, it gives perfect conditions for learning.

We will work with a very strict learnability criterion: finite identifiability—we will require that the output is a model that accurately describes the dependence, and it is obtained in finite time, with certainty. In the computational context this kind of learner is expressed as a Turing machine that, while receiving more and more data, at some finite step outputs the correct answer and then halts. This means that the moment of convergence to the right hypothesis is decidable.

To capture this in a more abstract way, we will here use the concept of one-shot learning—we require that for our problems there must be a learner that is *at most once defined*, meaning that for every stream $\varepsilon$, and any $n, k \in \mathbb{N}$ with $n \neq k$ either $\varepsilon[n] = \uparrow$ or $\varepsilon[k] = \uparrow$, and that the sole proper conjecture of the learner correctly describes he structure in question. This is technical characterisation of exact learning with certainty. Even though we could define some the intermediate, work-in-progress conjectures or best guesses, this learner, as long as it is uncertain, responds with $\uparrow$ (for discussion of the concept of finite identifiability and once-defined learners consult [10] and [8]).

**Definition 8.** *Given a simple causal frame* $\mathbb{C}$ *and a learner* $L$, *we say that* $L$ *finitely identifies* $\mathbf{D}^{\mathbb{C}}$ *on a stream* $\varepsilon$ *for* $\mathbb{C}$ *if* $L$ *is at most once-defined on* $\varepsilon$, *and there is an* $n \in \mathbb{N}$ *such that* $L(\varepsilon[n]) = \mathbf{D}^{\mathbb{C}}$. $L$ *finitely identifies* $\mathbf{D}^{\mathbb{C}}$ *if it finitely identifies* $\mathbf{D}^{\mathbb{C}}$ *on every stream for* $\mathbb{C}$.

In the remainder of this section, we will present two methods for finitely identifying dependence between variables. The first concerns the simple case when a variable depends on at most one other variable in the domain (single-variable causality), while the second allows variables to depend on multiple other variables (multi-variable causality).

### 4.1 Single-Variable Causality

Single-variable simple causality restricts our structures to only those causal frames in which every endogenous variable is causally influenced by exactly one exogenous variable. As we will see, this very simple setting allows us to lift the assumption that the agent know the difference between exogenous and endogenous variables at the start of the learning process.

We will first define a simple update function, which will work on a hypothesis space, $h_{dep} = \{\{x, y\} \mid \{x, y\} \subseteq U \cup V \; \& \; x \neq y\}$. Let $(s, t) \in S_{\mathcal{F}} \times S_{\mathcal{F}}$, we define $h_{dep} \upharpoonright (s, t) = \{\{x, y\} \mid s(x) = t(x) \text{ iff } s(y) = t(y)\}$ and, inductively, for $\sigma \in (S_{\mathcal{F}} \times S_{\mathcal{F}})^*$, $h_{dep} \upharpoonright \sigma \cdot (s, t) = (h_{dep} \upharpoonright \sigma) \upharpoonright (s, t)$.[1] The criteria for eliminating an element of $h_{dep}$ (see proposition 2) combined with the assumption of streams being sound and complete, means that all elements containing only exogenous variables will be eliminated during the learning process and we will thus allow the learner to build a hypothesis space with all possible pairs of variables to avoid an unnecessary pre-processing procedure or additional assumptions.

**Proposition 2.** *Let* $\mathbb{C}$ *be a simple single-variable causal frame and let* $\varepsilon$ *be a stream for* $\mathbb{C}$. $\mathbf{D}^{\mathbb{C}}$ *is finitely identified by the function:*

$$L_{dep}(\varepsilon[n]) = \begin{cases} h_{dep} \upharpoonright \varepsilon[n] & \text{if } |h_{dep} \upharpoonright \varepsilon[n]| = |V|, \\ \uparrow & \text{otherwise.} \end{cases}$$

---

[1] Here '·' stands for concatenation of sequences.

PROOF. Take a $\mathbb{C} = (U, V, \mathcal{F})$, and a stream $\varepsilon$ for $\mathbb{C}$. We need to show that for some $n$: 1) $|h_{dep} \upharpoonright \varepsilon[n]| = |V|$, and that 2) $\mathbf{D}^{\mathbb{C}} = (U, V, h_{dep} \upharpoonright \varepsilon[n])$. Take $n$ such that $\varepsilon[n]$ is a sequence for $\mathbb{C}$ (as specified in Definition 7). For 1), by the assumption of single-variable causality, every endogenous variable $v$ is influenced by exactly one exogenous variable $u$, so for all exogenous $u' \neq u$, $\varepsilon[n]$ will contain evidence of intervention on $u'$ that did not change the value of $v$, so such doubletons $\{u', v\}$ will have been eliminated from $h_{dep}$. Since this is the case for each $v \in V$, indeed $|h_{dep} \upharpoonright \varepsilon[n]| = |V|$. For 2), we need that each element $\{u, v\} \in h_{dep} \upharpoonright \varepsilon[n]$ contains all and only those exogenous variables that directly causally influence $v$. This is clearly the case. □

Let us apply this learning function to a simple example.

*Example 2.* We now consider a case where the action of pulling the blue lever has no effect, as represented in the state space in Fig. 2.

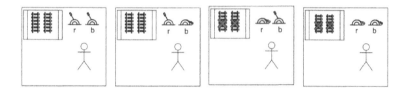

**Fig. 2.** The train track control domain

Formally, we have $U = \{r, b\}$, $V = \{t\}$, the propositions are presented in the following order $(r, b, t)$. The learner will start by building the hypothesis space from $U \cup V$, $h_{dep} = \{\{r, b\}, \{b, t\}, \{r, t\}\}$. The stream of observations starts with $\varepsilon[0] = ((0, 0, 1), (0, 1, 1))$, which corresponds with intervention $b := 1$ (pulling down the blue lever). Then, $h_{dep} \upharpoonright \varepsilon_0 = \{\{r, b\}, \{b, t\}, \{r, t\}\}$.

After observing $\varepsilon_0$, the learner has correctly identified the dependence $\{r, t\}$.[2]

## 4.2 Multi-variable Causality

Let us now consider the case where some endogenous propositions depend on multiple exogenous propositions. First we will show that the method from the previous subsection will not suffice.

*Example 3.* Recall the train track control example from Fig. 1; in order to merge the tracks, both levers must be pulled. The hypothesis space according to the previously defined method would be:

$$h_{dep} = \{\{r, b\}, \{b, t\}, \{r, t\}, \{r, m\}, \{b, m\}, \{t, m\}\}.$$

---

[2] Note that the same result is obtained by the learner for any sound and complete stream for the causal frame of this particular domain.

Let us fix the order of propositions as $(r, b, t, m)$. Let the start of the stream be: $\varepsilon_0 = ((1,0,0,0),(0,0,1,0))$, and $\varepsilon_1 = ((0,1,1,0),(1,1,0,1))$. The first transition results from the intervention $r := 0$ and the second one from $r := 1$. The learner proceed as follows:

$$h_{dep}\!\restriction\!\varepsilon_0 = \{\,\{r,b\},\ \{b,t\},\ \{r,t\},\ \{r,m\},\ \{b,m\},\ \{t,m\}\,\}.$$

$$h_{dep}\!\restriction\!\varepsilon_1 = \{\,\{r,b\},\ \{b,t\},\ \{r,t\},\ \{r,m\},\ \{b,m\},\ \{t,m\}\,\}.$$

Here our learner is only able to identify the dependence between $\{r, t\}$, as all possible pairs including $b$ are ruled out by the second observation. This clearly shows that our learner is unable to identify dependencies where one proposition in our domain is dependent on multiple other propositions.

As shown in Example 3 our learning function works by eliminating possible dependencies. We could extend the hypothesis space to include all possible combinations of dependence in $U \cup V$, but this becomes a rather tedious process when $U \cup V$ is large. We will therefore consider an additive approach to our learning problem. We will now assume that our learner distinguishes between $U$ and $V$, that the manipulable variables are known to the learner, and that they coincide with the set $U$. The learning algorithm for learning multi-variable dependencies (Algorithm 1) will use this knowledge.

---

**Algorithm 1** $v_i-\mathrm{Dependence}(v_i, k, \varepsilon[\ell])$

---

    **Input** $v_i$ (endogenous variable),
    **Input** $k$ (is the number of exogenous variables),
        $\varepsilon[\ell]$ (finite sequence for $\mathbb{C}$)
    **Output** $f$ (dependence set of $v_i$)
1: $f = \emptyset$
2: **for** $j = 0, \ldots, \ell$ **do**
3:     $s$ = first element of $\varepsilon_j$
4:     $s'$ = second element of $\varepsilon_j$
5:     **if** $(n + i \in s \# s')$ **then**
6:         $f.add(s \# s')$
7: $f.remove(\{\ell \mid \ell > n\})$
8: $f.add(n + i)$
9: **return** $f$

---

Given the set $V$ with $|V| = k$, a stream $\varepsilon$ for $\mathbb{C}$, and $n \in \mathbb{N}$, the learner will, for each $v \in V$ identify the dependence set $F_v$ of $\mathbf{D}^{\mathbb{C}}$. This is done using the $v_i-\mathrm{Dependence}(v_i, k, \varepsilon[n])$ procedure shown above. When each individual $v \in V$ has been investigated, the learner can build the full set $F$ of $\mathbf{D}^{\mathbb{C}}$.

Let us briefly explain the pseudo-code Algorithm 1. First let us fix the enumeration of all variables starting with all $n$ exogenous variables and following with all $k$ endogenous variables $\{x_0, \ldots, x_{n-1}, x_n, \ldots, x_{n+k-1}\}$. The algorithm constructs the dependence set for a given endogenous variable $v_i$, that in this enumeration has the index $n + i$. For each observation of intervention $(s, s')$ in $\epsilon[\ell]$ it computes the set of indices of variables that changed their value during this intervention, i.e., the set $s \# s'$. If the index of our variable $v_i$ (i.e., $n + i$) is in that set, we add to our dependence set all indices in $s \# s'$. After this has been performed for all $\ell$ steps, the index of $v_i$ (i.e., $n + i$) is added to the dependence set.

Let us now apply the $v_i$-Dependence on the train track control Example 3. The elements of the given observation stream $\varepsilon[4]$ are shown in each of the Tables 1 and 2, and they result from the following interventions: $r := 0$, $r := 1$, $b := 1$. We order the propositions in the following way: $(r, b, t, m)$, $k = 2$. The procedure of $v_i-\text{Dependence}(m, 2, \varepsilon[4])$, is shown in Table 1, and of $v_i-\text{Dependence}(t, 2, \varepsilon[4])$ in Table 2. The outcome of $v_i-\text{Dependence}(v_i, k, \varepsilon[n])$ on both endogenous propositions of the domain $t$ and $m$, is then the two dependencies $\{r, t\}$ and $\{r, b, m\}$, which can be expressed in DL as $=(r, t)$ and $=(r, b, t)$.

**Table 1.** $v_i-\text{Dependence}(m, 2, \varepsilon[4])$.

| $j$ | $\varepsilon_j$ | $f_m$ |
|---|---|---|
| 0 | $((1,0,0,0),(0,0,1,0))$ | $\emptyset$ |
| 1 | $((0,1,1,0),(1,1,0,1))$ | $\{r, m\}$ |
| 2 | $((1,0,0,0),(1,1,0,1))$ | $\{r, b, m\}$ |
| 3 | $((0,0,1,0),(0,1,1,0))$ | $\{r, b, m\}$ |

**Table 2.** $v_i-\text{Dependence}(t, 2, \varepsilon[4])$.

| $j$ | $\varepsilon_j$ | $f_t$ |
|---|---|---|
| 0 | $((1,0,0,0),(0,0,1,0))$ | $\{r, t\}$ |
| 1 | $((0,1,1,0),(1,1,0,1))$ | $\{r, t\}$ |
| 2 | $((1,0,0,0),(1,1,0,1))$ | $\{r, t\}$ |
| 3 | $((0,0,1,0),(0,1,1,0))$ | $\{r, t\}$ |

**Theorem 1.** *Let* $\mathbb{C} = (U, V, \mathcal{F})$ *be a simple multi-variable causal frame,* $v_i \in V$, $|V| = k$, *and* $\varepsilon[n]$ *a sequence for* $\mathbb{C}$. $v_i-\text{Dependence}(v_i, k, \varepsilon[n])$ *outputs the set* $F_{v_i}$ *in* $\mathbf{D}^{\mathbb{C}}$.

PROOF. Let $\mathbf{D}^{\mathbb{C}} = (U, V, F)$, $F_v \in F$, and $\varepsilon[n]$ be a sequence for $\mathbb{C}$. Assume that the procedure executed for $v$ outputs $F'$, we need to show that $F' = F_v$, i.e., for all $x \in U \cup V$, $x \in F'$ iff $x \in F_v$. For $x \in V$, it is clearly the case since the only endogenous variables in $F'$ (by line 7 and 8 in Algorithm 1) and $F_v$ (by definition of DMs) is $v$. It remains to show that for all $x \in U$, $x \in F'$ iff $x \in F_v$. If $x \in F'$, then there is an intervention $(s, s')$ in $\varepsilon[n]$, such that the index of $v$ is in $s \# s'$ and the index of $x$ is in $s \# s'$, which means there is a valuation on $U$ such that the intervention (solely) on $x$ changes the value of $v$. So, $x$ directly causally influences $v$, so $x \in F_v$. For the other direction, assume that $x \in F_v$, which means that $x$ directly causally influences $v$ in $\mathbb{C}$. Since $\varepsilon[n]$ is for $\mathbb{C}$, it includes an observation $\varepsilon_\ell$ (for some $\ell < n$) of an intervention that supports that fact. At that iteration $\ell$, $x$ is added to $F'$ (lines 5, 6 in Algorithm 1).  □

**Corollary 1.** *If* $\mathbb{C}$ *be a simple multi-variable causal frame, then* $\mathbf{D}^{\mathbb{C}}$ *is finitely identifiable.*

PROOF. As $v_i$−Dependence$(v_i, k, \varepsilon[n])$ correctly identifies the dependence set of each $v_i \in V$, using this algorithm for all $v \in V$ will identify the dependencies of $\mathbf{D}^C$.                                                                                                    □

# 5   Complexity

Dependence models can be viewed as (non-lossless) compression of causal frames. There are $2^{|U \cup V|} - 1$ possible non-empty combinations of variables $U$ and $V$, which is an upper bound on the number of possible $F$s, i.e., possible dependence models over $U \cup V$. On top of that causal frames will allow for each combination $2^{|U \cup V|}$ (binary) valuations over $U$ and $V$, giving a $2^{|U \cup V| \cdot 2^{|U \cup V|}} - 1$ of possible $\mathcal{F}$s. This compression is a vast improvement in the memory needed to represent the structure of the causal relations. As cognitive and artificial agents have very limited memory, the dependence models seem to be a more likely way in which information about causation is stored. Arguably, in many natural scenarios it is enough to know which buttons and switches control which lamps, the exact relationship between their configurations is less important for efficient interaction with the environment.

Let us consider the time and space complexity of our learning procedures: $L_{dep}(\varepsilon[n])$ and the $v_i$− Dependence$(v_i, k, \varepsilon[n])$ learner. $L_{dep}(\varepsilon[n])$ is defined in terms of the hypothesis space $h_{dep}$ update procedure. We will therefore start by analyzing $h_{dep}$. As the space complexity of $h_{dep}$ is upper-bounded by the number of pairs in $U \cup V$, $\binom{|U \cup V|}{2}$, $O(\binom{|U \cup V|}{2})$ is the space complexity of $L_{dep}(\varepsilon[n])$ as well. To analyze the time complexity of $L_{dep}$, we could either choose to ignore repetitions in $\varepsilon[n]$ or assume that all observations are distinct. In either case, the time complexity of $L_{dep}$ will be defined in terms of the number of distinct pairs of observations in $\varepsilon[n]$ the learner must receive to learn $F$. In order to exclude all possible dependencies in $U \cup V$ except $F$, at most $|U \cup V| - 2$ distinct observations are needed, as the learner will require to see all propositions not in $F$ change value independently in order to be certain if the dependence is indeed $F$. This gives $L_{dep}(\varepsilon[n])$ time complexity of $O(|U \cup V|)$.

For the $v_i$−Dependence$(v_i, k, \varepsilon[n])$ learner, the space complexity for every $v_i \in V$ is $O(|U \cup V|)$, as the learner will in the worst case not be able to add propositions to the $v_i$−Dependence$(v_i, k, \varepsilon[n])$. For the total set of dependencies $V$ the learner will have a space complexity of $O(|U \cup V|^2)$. The time of $v_i$−Dependence$(v_i, k, \varepsilon[n])$ complexity will also be $O(|U \cup V|)$ for each $v \in V$, as this is the maximum number of distinct observation pairs in $\varepsilon[n]$ the learner needs to learn the dependence of $v$. In total the time complexity will thus be $O(|U \cup V|^2)$.

# 6   Conclusion and Discussion

In this paper we have set the stage for a clear and comprehensive framework of learning by intervention in causal frames. We have proposed two learning methods: a learning function to handle finite identifiability of single-variable simple dependence and a learning algorithm to handle a more general multi-variable notion of dependence. We have presented two proposals for an exact learner of cause-effect relations in graphical models, in the style of [12], in recent years combined with epistemic modal logic (for an overview, see [9]). This departs from the traditional probabilistic learning methods as shown in the overviews of [11] and [19]. One among many motivations for an exact learners is that we for some environments it may be extremely costly or impossible to obtain the amount of data needed for statistical models to perform well, and thus we need some qualitative methods to discover the causal relations in such environment.

*Related Work.* An interesting connection is that with logics of dependence. As our learners infer dependencies between variables without uncovering the exact way in which variables causally influence each other, they can be seen as ways to learn dependence atoms. Interestingly, team semantics has recently been applied to describe interventionist counterfactuals and causal dependencies in [3]. Introduced therein *causal teams* bear resemblance to our dependence models. This connection is worth pursuing further.

Since our learners converge to *knowledge* of a certain causal structure, a natural question is how our setup relates to the recently developed epistemic logic of causality $\mathcal{L}_{PAKC}$ [4]. The language $\mathcal{L}_{PAKC}$ contains expressions '$X{=}x$' for interventions (the variable $X$ has value $x$), '$[X{=}x!]$' for announcements (or observations) of interventions, and a knowledge operator '$K$'. $\mathcal{L}_{PAKC}$ is interpreted over epistemic causal models—the following is a version of that concept suited to our *simple* causal models:

**Definition 9 (Simple Epistemic Causal Model).** *A simple epistemic causal model over a domain is a tuple* $\mathbf{E} = (U, V, \mathcal{F}, \mathcal{T})$, *where* $U$, $V$, *and* $\mathcal{F}$ *are as in simple causal models;* $\mathcal{T}$ *is a non-empty set of valuations complying with* $\mathcal{F}$.

The uncertainty of the agent in an epistemic causal model of [4] ranges only over valuations that comply to the set of structural functions $\mathcal{F}$, which means that the agent's knowledge always accounts for the true causal structure of the model. However useful this restriction might be for constructing a sound and complete logic of interventions [4], it does not fare well with our learning scenario. To model an agent learning an unknown domain we must allow that her uncertainty (at least initially) ranges also over valuations that might not comply with $\mathcal{F}$. In the sequel of the present paper we want to extend the framework of [4], to allow for the uncertainty of the agent to include valuations that do not comply to a given set of structural functions (as is especially clear in the case of $L_{dep}(\epsilon[n])$ in Sect. 4.1). Our learning condition could then be expressed with the use of this new language in a way similar to that in which learnability is expressed in Dynamic Logic for Learning Theory [2]: given a causal frame, starting in a given model there is a sequence of interventions after which the agent knows the underlying causal structure, i.e., for all variables the agent knows which are directly causally related to each other (with the use of $\rightsquigarrow$ operator in $\mathcal{L}_{PAKC}$ [4]).

*Possible Extensions.* The directions for further work are numerous. The first group of topics concerns the relaxation of our simplifying assumptions, and tackling the full complexity of causal frames in the context of learning. Our current methods will identify the topologies show in Fig. 3. Due to the restriction that only exogenous variables can causally influence endogenous variables, chains (of at least length two) and confounders will not be identified by our learners. It would therefore be a natural next step to extend the methods to include these two topologies as well as other more complex causal relations. One of the challenges to achieve this is to enable the learner to distinguish between a chain and a fork given a sound and complete observation stream, which would impose further restrictions on what it would be required of a stream to be sound and complete for a given dependence. Another extension would be to allow our variables to be non-binary, thus bringing us closer to real-world cause-effect relations. This relaxation should not add much complexity to our dependence learners, as they are not concerned with the exact values of variables in a causal relation, but simply the *existence* of a causal relation between variables. The main addition to the existing algorithm will be an update of the completeness criteria of the input stream, it must contain all possible valuations of the endogenous variables to allow the dependence learner to eliminate relations to the exogenous variables.

**Fig. 3.** Topologies of simple causal models; a chain of length 1, collider, fork, mediator

Moreover, so far we assumed that the set of manipulable variables coincides with the set of exogenous variables. It could be interesting to investigate the case where $M \subset U$, and thus only some exogenous variables can be manipulated by the agent. Next, we could relax the condition of full-observability. As most real-world problems involve partial observability of the world, adapting our model to handle this as well would be a natural next step [5]. Another assumption of this paper is that the domain is static, which is a simplification with respect to many real-world problems where things constantly change. Applying algorithms known from dynamic graph theory might provide inspiration on how agents can learn dependence efficiently in unknown dynamic domains.

In this paper, we have shown how agents can learn global dependence between propositions in an unknown domain, as defined in [1]. It would therefore be interesting to investigate the perspective of local dependence model learning, either to provide a subroutine for finding global dependence or as an independent study of causation. Another approach is to look further into the properties of the causal models presented in the Halpern-Pearl Actual Causality [13], where from the set of structural functions a graphical representation of the causal structures can be build, which provides the agent with a visual representation of the underlying causalities in their domain. We would like to check if such a representation can be beneficial to a learner. The possibility of adding probabilities to these causal networks could be another interesting approach to investigate the prediction level of Pearl's Ladder.

# References

1. Baltag, A., van Benthem, J.: A simple logic of functional dependence. J. Philos. Log. **50**(5), 939–1005 (2021). https://doi.org/10.1007/s10992-020-09588-z
2. Baltag, A., Gierasimczuk, N., Özgün, A., Vargas Sandoval, A.L., Smets, S.: A dynamic logic for learning theory. J. Log. Algebraic Methods Program. **109**, 100485 (2019). https://doi.org/10.1016/j.jlamp.2019.100485
3. Barbero, F., Sandu, G.: Team semantics for interventionist counterfactuals: observations vs. interventions. J. Philos. Log. **50**(3), 471–521 (2021). https://doi.org/10.1007/s10992-020-09573-6
4. Barbero, F., Schulz, K., Smets, S., Velázquez-Quesada, F.R., Xie, K.: Thinking about causation: a causal language with epistemic operators. In: Martins, M.A., Sedlár, I. (eds.) DaLi 2020. LNCS, vol. 12569, pp. 17–32. Springer, Cham (2020). https://doi.org/10.1007/978-3-030-65840-3_2
5. Bolander, T., Gierasimczuk, N., Liberman, A.: Learning to act and observe in partially observable domains. In: Bezhanishvili, N., Yang, F. (eds.) Outstanding Contributions to Logic: Dick de Jongh. Springer, Cham (2023). (in preparation). https://doi.org/10.48550/arXiv.2109.06076
6. Bolander, T., Gierasimczuk, N.: Learning actions models: qualitative approach. In: van der Hoek, W., Holliday, W.H., Wang, W. (eds.) LORI 2015. LNCS, vol. 9394, pp. 40–52. Springer, Heidelberg (2015). https://doi.org/10.1007/978-3-662-48561-3_4
7. Bolander, T., Gierasimczuk, N.: Learning to act: qualitative learning of deterministic action models. J. Log. Comput. **28**(2), 337–365 (2017). https://doi.org/10.1093/logcom/exx036
8. Dégremont, C., Gierasimczuk, N.: Finite identification from the viewpoint of epistemic update. Inf. Comput. **209**(3), 383–396 (2011). https://doi.org/10.1016/j.ic.2010.08.002
9. Gierasimczuk, N.: Inductive inference and epistemic modal logic. In: Klin, B., Pimentel, E. (eds.) 31st EACSL Annual Conference on Computer Science Logic (CSL 2023). Leibniz International Proceedings in Informatics (LIPIcs), vol. 252, pp. 2:1–2:16. Schloss Dagstuhl - Leibniz-Zentrum für Informatik, Dagstuhl (2023). https://doi.org/10.4230/LIPIcs.CSL.2023.2
10. Gierasimczuk, N., Jonghde Jongh, D.: On the complexity of conclusive update. Comput. J. **56**(3), 365–377 (2013). https://doi.org/10.1093/comjnl/bxs059
11. Glymour, C., Zhang, K., Spirtes, P.: Review of causal discovery methods based on graphical models. Front. Genet. **10**, 524 (2019). https://doi.org/10.3389/fgene.2019.00524
12. Gold, M.: Language identification in the limit. Inf. Control **10**(5), 447–474 (1967). https://doi.org/10.1016/S0019-9958(67)91165-5
13. Halpern, J.Y.: Actual Causality. The MIT Press, Cambridge (2016). https://doi.org/10.7551/mitpress/10809.001.0001

14. Hintikka, J., Sandu, G.: Informational independence as a semantical phenomenon. In: Fenstad, J.E., Frolov, I.T., Hilpinen, R. (eds.) Logic, Methodology and Philosophy of Science VIII, Studies in Logic and the Foundations of Mathematics, vol. 126, pp. 571–589. Elsevier (1989). https://doi.org/10.1016/S0049-237X(08)70066-1

15. McCormack, T., Bramley, N., Frosch, C., Patrick, F., Lagnado, D.: Children's use of interventions to learn causal structure. J. Exp. Child Psychol. **141**, 1–22 (2016). https://doi.org/10.1016/j.jecp.2015.06.017

16. Muentener, P., Bonawitz, E.: The development of causal reasoning. In: Waldmann, M.R. (ed.) The Oxford Handbook of Causal Reasoning, pp. 677–698. Oxford University Press (2017). https://doi.org/10.1093/oxfordhb/9780199399550.001.0001

17. Pearl, J.: Introduction to Probabilities, Graphs, and Causal Models, 2 edn., pp. 1–40. Cambridge University Press (2009). https://doi.org/10.1017/CBO9780511803161.003

18. Pearl, J., Mackenzie, D.: The Book of Why: The New Science of Cause and Effect, 1st edn. Basic Books Inc., USA (2018). https://doi.org/10.1007/s00146-020-00971-7

19. Peters, J.M., Janzing, D., Schölkopf, B.: Elements of Causal Inference: Foundations and Learning Algorithms: Foundations and Learning Algorithms. MIT Press, Cambridge (2017)

20. Väänänen, J.: Dependence Logic: A New Approach to Independence Friendly Logic (London Mathematical Society Student Texts). Cambridge University Press, Cambridge (2007). https://doi.org/10.1017/CBO9780511611193

21. Von Wright, G.: On the logic and epistemology of the causal relation. In: Suppes, P., Henkin, L., Joja, A., Moisil, G.C. (eds.) Proceedings of the Fourth International Congress for Logic, Methodology and Philosophy of Science, Bucharest, 1971. Studies in Logic and the Foundations of Mathematics, vol. 74, pp. 293–312. Elsevier (1973). https://doi.org/10.1016/S0049-237X(09)70366-0

22. Yang, F., Väänänen, J.: Propositional logics of dependence. Ann. Pure Appl. Log. **167**(7), 557–589 (2016). https://doi.org/10.1016/j.apal.2016.03.003

# A Logical Approach to Doxastic Causal Reasoning

Kaibo Xie[1], Qingyu He[2(✉)], and Fenrong Liu[2]

[1] Wuhan University, Wuhan, China
xiekaibozju@gmail.com
[2] Tsinghua University, Beijing, China
qingyuhethu@gmail.com, fenrong@tsinghua.edu.cn

**Abstract.** Belief revision and causality play an important role in many applications, typically, in the study of database update mechanisms and data dependence. New contributions on causal reasoning are continuously added to the pioneering works by Pearl, Halpern and others. Though there is a long tradition of modeling belief revision in philosophical logic, the entanglement between belief revision and causal reasoning has not yet been fully studied from a logical view. In this paper, we propose a new formal logic for doxastic causal reasoning. With examples, we illustrate that our framework explains the rational way of belief revision based on causal reasoning. We further study the general properties of the logic. A complete axiomatization, as well as a decidability result, will be given. In addition, we believe our work will shed light on understanding the relation between qualitative and quantitative approaches toward (causal) dependence in general.

**Keywords:** Causal reasoning · Plausibility model · Causal model · Conditional Doxastic logic

## 1 Introduction

How to characterize the mechanisms of agents' belief change is an important question in the tradition of philosophy. In recent years, a lot of work, such as [4,6] characterizes doxastic reasoning from a logical point of view. This research has already gone beyond the borders of philosophy and has various applications in computer science and artificial intelligence(see, e.g. [8,16]). In those areas, there are many examples of belief revision that involve the entanglement between doxastic reasoning and causal reasoning. For instance, when an agent's belief is revised with a new proposition $P$, she should preserve her beliefs about those facts that are causally independent of $P$. Yet there is no account of belief revision in the literature that explicitly takes causal reasoning into account. This paper will propose a formal framework to characterize belief revision based on causality.

The entanglement between doxastic reasoning and causal reasoning can be best illustrated in the following example:

Supported by Tsinghua University Initiative Scientific Research Program.

**Example 1.** *If John has a talent for science (T=1), then it is very probable that he excels in both chemistry (C=1) and physics (P=1). Furthermore, given that a college of science and engineering tends to prioritize applicants who are good at chemistry or physics, it is very probable that John would be accepted into the college (A=1).*

The causal structure of this example can be represented by the graph below:

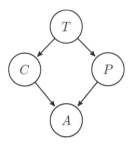

Now, let us consider the independence of belief in this example. Intuitively, the belief that John excels in physics is dependent on the belief that John excels in chemistry. Given the information that John excels in physics, the agent is more likely to believe that John excels in chemistry. The dependence results from both doxastic reasoning and causal reasoning: John's talent for science $(T)$ is a "significant" cause of John's excellent ability in chemistry $(C)$, given $C$, $T$ is very likely to be true.

Information update also plays an important role in the dependence of belief. Suppose the agent is informed that John has no talent for science, then knowing $C$ will not increase the likelihood of $P$. So the information update of $T$ breaks the dependence of belief between $C$ and $P$. Information updates can not only break the dependence but also build the dependence. For instance, if the agent is further informed that John is accepted by the college, then the agent tends to believe that John excels in chemistry once given John is not good at physics.

As illustrated in Example 1, there is an intriguing entanglement between the epistemic perspective and causal considerations in the agent's reasoning: the information update on the only common cause of two variables breaks the dependence of belief between them; In contrast, the information update on the common consequence creates new dependence of beliefs.

We are aware that in recent years a lot of effort has been put into the study of probabilistic causal reasoning, such as the well-known causal Bayesian network developed in [18,20], and the semi-deterministic probabilistic causal model proposed in [12]. In order to merge causal reasoning with doxastic reasoning, we will propose a model which embeds the causal structure into an epistemic model. In addition, we will develop a logic framework that not only captures the properties of belief revision but also reflect the natural features of probabilistic causal reasoning.

The rest of the paper is organized as follows. Section 2 is a brief review of the theories of belief and causality. In Sect. 3, we introduce the model for doxastic causal reasoning. In Sect. 4, we give the syntax and semantics of the language of doxastic causal reasoning and define some interesting notions of causality. In Sect. 5, we discuss the correspondence between quantitative and qualitative approaches. In Sect. 6, we present the complete axiomatization and the result of decidability of this logic. We conclude this paper in Sect. 7.

## 2   Formal Representation of Belief and Causality

In order to account for the reasoning of belief based on causal information (as in Example 1), we will build our work upon the latest research in both fields, the theories of belief revision and causal reasoning. In this section, we present the necessary building blocks.

In the literature of epistemic logic, the knowledge of an agent is represented by the notion of "epistemic distinguishability": for each possible world $w$, there is a set $s(w)$ which consists of all the worlds which the agent cannot distinguish at $w$. An agent knows $\phi$ at $w$ whenever $\phi$ holds at every world in $s(w)$. Many logicians define belief in terms of "epistemic distinguishability" together with the notion of "plausibility": an agent believes $\phi$ at $w$ whenever $\phi$ holds at the most plausible epistemic indistinguishable worlds. [4,7] proposed a plausibility model which formalizes both "epistemic distinguishability" and "plausibility". The language of conditional doxastic logic expresses the epistemic state of an agent by epistemic operators. $K\phi$ stands for "the agent knows that $\phi$", $B\phi$ stands for "the agent believes that $\phi$" and "$B^\psi\phi$" stands for "the agent believes $\phi$ conditional on $\psi$". Based on the plausibility model, the conditional doxastic logic defines the truth condition of conditional belief $B^\psi\phi$ as: $\phi$ holds at the most plausible epistemic indistinguishable world where $\psi$ holds. We embrace a qualitative representation of belief within our framework, however belief can also be defined through subjective probability. The connection between quantitative and qualitative belief representations is studied by [14,15].

Next, to represent a causal structure, we will make use of the structural equation model developed in [9,17]. Intuitively, a causal structure consists of two parts: the causal variables and the causal influence among those variables. Formally, the causal variables can be described by a signature $\mathcal{S}=(\mathcal{U},\mathcal{V},\Sigma)$ where $\mathcal{U}$ is a finite set[1] of exogenous variables, $\mathcal{V}$ is a finite set of endogenous variables, $\Sigma$ is the range of the variables. In a structural equation model, the causal influence among causal variables is usually represented by a set of structural functions $\mathcal{F}$: for each endogenous variable $X$, $\mathcal{F}$ contains a function $f_X$ which tells the value of $X$ given all of the other variables. Formally, a causal model is defined as a tuple $\langle \mathcal{S}, \mathcal{F} \rangle$ where $\mathcal{S} = (\mathcal{U}, \mathcal{V}, \Sigma)$ is the signature, $\mathcal{F}$ is a collection of functions $\{f_X\}_{X \in \mathcal{V}}$ with $f_X : ((\mathcal{U} \cup \mathcal{V})\setminus\{X\} \to \Sigma) \to \Sigma$. $f_X$ is called a structural equation function of $X$. In many studies of structural equation models, a causal model is

---

[1] The investigation into causal models with infinite variables is presented within [11].

usually assumed to be *acyclic* or *recursive*, which intuitively means the causal influence represented by $\mathcal{F}$ is acyclic[2].

The structural equation model can be used to define the notion of intervention. Intervention is a hypothetical change of the actual state (as well as the causal rules) which forces the value of some (endogenous) variables to be changed. Let $\langle \mathcal{S}, \mathcal{F} \rangle$ be a causal model and let $\mathcal{A}$ be an assignment to variables representing the actual state. The result of an intervention is defined as below.

**Definition 1.** *The causal model results from an intervention forcing the value of $\vec{X}$ to be $\vec{x}$ is defined as $\langle \mathcal{S}, \mathcal{F}_{\vec{X}=\vec{x}} \rangle$ and the actual value after the intervention is defined as $\mathcal{A}^{\mathcal{F}}_{\vec{X}=\vec{x}}$ where:*

- *the functions in $\mathcal{F}_{\vec{X}=\vec{x}} = \{ f'_V \mid V \in \mathcal{V} \}$ are such that: (i) for each $V$ **not in** $\vec{X}$, the function $f'_V$ is exactly as $f_V$, and (ii) for each $V = X_i \in \vec{X}$, the function $f'_{X_i}$ is a constant function returning the value $x_i \in \vec{x}$ regardless of the values of all other variables.*
- *$\mathcal{A}^{\mathcal{F}}_{\vec{X}=\vec{x}}$ is the unique[3] assignment to $\mathcal{F}_{\vec{X}=\vec{x}}$ whose assignment to exogenous variables is identical with $\mathcal{A}$. Formally, $\mathcal{A}^{\mathcal{F}}_{\vec{X}=\vec{x}}(Y)$ is the unique assignment that satisfies the following equations:*

$$\mathcal{A}^{\mathcal{F}}_{\vec{X}=\vec{x}}(Y) = \begin{cases} \mathcal{A}(Y) & if\, Y \in \mathcal{U} \\ f'_Y((\mathcal{A}^{\mathcal{F}}_{\vec{X}=\vec{x}})^{-Y}) & if\, Y \in \mathcal{V}. \end{cases}$$

*Note that $(\mathcal{A})^{-X}$ denotes the sub-assignment of $\mathcal{A}$ to $(\mathcal{U} \cup \mathcal{V}) \backslash \{X\}$.*

Although the structural equation functions are *deterministic* (as the value of a variable is determined when given the value of all other variables), they are also able to represent non-deterministic causal influence. The causal modelling approach interprets non-deterministic causal influence in a "Laplacian way". According to the Laplacian interpretation of causal influence, the non-deterministic causal relation between John's talent and his ability in chemistry in Example 1 can be explained as follows: there is some variable $U_C$ which represents all the unknown possible factors that influence the realization of John's talent (for example, John is more interested in playing video games than attending classes.). For instance, the structural function for $C$ can be defined as: $\mathcal{F}_C(\mathcal{A}^-) = 1$ whenever $T = 1$ and $U_C = 1$, where $\mathcal{A}^-$ is a partial assignment to all variables other than $C$. Thus the randomness of the value of $C$ given $T$ is reduced to the ignorance of the value of $U_C$.

Following the Laplacian interpretation of randomness, the causal structure in Example 1 can be represented by a structural equation model $\langle \mathcal{S}, \mathcal{F} \rangle$, which can be graphically represented as Fig. 1.

---

[2] Namely there is no sequence $X_1, ..., X_n$ such that for each $0 < k < n$ the value of $X_{k+1}$ is dependent on $X_k$ according to $\mathcal{F}$, and the value of $X_1$ is also dependent on $X_n$.

[3] Since $\mathcal{F}$ is acyclic, $\mathcal{F}_{\vec{X}=\vec{x}}$ is also acyclic. Thus $\mathcal{F}_{\vec{X}=\vec{x}}$ has a unique solution with respect to each setting of exogenous variables.

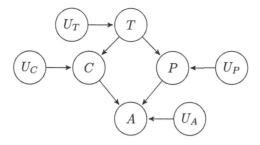

**Fig. 1.** The graphical representation of the "Laplacian causal model" for Example 1

The variables in $\mathcal{S}$ are nodes of the graph. The structural function in $\mathcal{F}$ are defined in a way that for each variable $V$, only those variables that are $V$'s parents in Fig. 1 matter in $\mathcal{F}_V$. Thus, we have an intuitive way to represent the causal structure of Example 1. We will come back to the example later.

## 3  Combining Belief and Causality

In this section, we will combine the two approaches above and propose a causal plausibility model which formalizes the reasoning of belief based on causal knowledge. Our work is partially inspired by the very recent proposals made in [5,13]. Specifically, the key idea of these proposals is that we can think of each possible assignment to all causal variables as a state/world in the plausibility model. The causal knowledge of an agent is represented by those possible worlds that comply with the causal rules. Formally, according [5], let the causal rules be represented by the set of structural equations $\mathcal{F}$, the set of possible worlds that complies with the causal rules can be represented by the set of assignments $W^{\mathcal{F}}$, where $W^{\mathcal{F}} = \{\mathcal{A} \in \Sigma^{\mathcal{U} \cup \mathcal{V}} \mid \forall X \in \mathcal{V}, \mathcal{A}(X) = f_X((\mathcal{A})^{-X})\}$.

Based on the formal representation of knowledge and belief introduced in Sect. 2, we will generalize this approach from causal knowledge to causal belief. The causal plausibility model proposed here is a plausibility model extended with a causal structure $\langle \mathcal{S}, \mathcal{F} \rangle$, whose plausibility relation is a binary relation over $W^{\mathcal{F}}$. Let us first define the basic causal plausibility model as follows:

**Definition 2 (Basic Causal Plausibility Model).** *A **basic causal plausibility model** $M$ is a tuple $M = \langle \mathcal{S}, \mathcal{F}, \leq, \mathcal{A} \rangle$ where:*

- *$\mathcal{S} = (\mathcal{U}, \mathcal{V}, \Sigma)$ is the signature (as in a structural equation model).*
- *$\mathcal{F}$ is a set of structural functions $\{f_X\}_{X \in \mathcal{V}}$ with $f_X : ((\mathcal{U} \cup \mathcal{V}) \setminus \{X\} \to \Sigma) \to \Sigma$. $\mathcal{F}$ is assumed to be acyclic.*
- *$\leq$ is a total order over $W^{\mathcal{F}}$.*
- *$\mathcal{A}$ is an assignment in $W^{\mathcal{F}}$.*

In this model $\mathcal{S}, \mathcal{F}$ represents the causal structure; $\leq$ represents the plausibility ordering of the agent to the value of causal variables; $\mathcal{A}$ represents the actual value of the causal variables.

We call the model defined in Definition 2 "basic" because we do not impose any further restriction on the plausibility ordering, as long as the ordering is over $W^{\mathcal{F}}$. However, this may be too arbitrary: let $U_1$ and $U_2$ be two exogenous variables, it could be the case that all of the most plausible worlds in $W^{\mathcal{F}}$ assign $U_1$ with value 1 but all of the most plausible worlds in which $U_2 = 1$ assign $U_1$ with value 0. By the classical interpretation of conditional belief, this intuitively means that the agent changes the belief about $U_2$ conditional on the information about $U_1$. However, this is irrational because exogenous variables are assumed to be causally independent and the agent is assumed to have this causal knowledge. Therefore, a rational agent's belief about exogenous variables should be independent according to our model (if there is no additional information), and the plausibility ordering should reflect this feature as well.

Based on this consideration, we propose the following restriction on the plausibility ordering in a causal plausibility model:

**Definition 3.** *The plausibility ordering in* $\langle \mathcal{S}, \mathcal{F}, \leq, \mathcal{A} \rangle$ *is* **uniform** *when the following property holds:*

*For any* $\mathcal{A}_1, \mathcal{A}_2 \in W^{\mathcal{F}}$ *and* $\vec{U} \in \mathcal{U}$, $\{\vec{U^-}\} = \mathcal{U} \backslash \{\vec{U}\}$, *if* $\mathcal{A}_1 \leq \mathcal{A}_2$ $\mathcal{A}_1(\vec{U^-}) = \mathcal{A}_2(\vec{U^-})$, *then for any* $\mathcal{A}_1'$ *and* $\mathcal{A}_2'$, $\mathcal{A}_1'(\vec{U}) = \mathcal{A}_1(\vec{U})$, $\mathcal{A}_2'(\vec{U}) = \mathcal{A}_2(\vec{U})$ *and* $\mathcal{A}_1'(\vec{U^-}) = \mathcal{A}_2'(\vec{U^-})$ *implies* $\mathcal{A}_1' \leq \mathcal{A}_2'$.

This restriction means that the plausibility ordering between two settings of exogenous variables is invariant under uniformly changing the value of any exogenous variables. The condition of uniformity intuitively expresses the independence among exogenous variables in belief.

A *uniform causal plausibility model* is a basic causal plausibility model whose plausibility ordering is uniform. For simplicity, in the rest of the paper when we say causal plausibility model, we mean it is a uniform causal plausibility model.

# 4 The Logic of Doxastic Causal Reasoning

## 4.1 Syntax and Semantics

Since the model we proposed in Sect. 3 integrates both the causal and plausibility models, in this section, we introduce a formal language to talk about knowledge, belief, and causation. The formal language combines the language of conditional doxastic logic and the logic for causal reasoning.

**Definition 4 (Language for doxastic causal reasoning).** *Let* $\mathcal{S} = (\mathcal{U}, \mathcal{V}, \Sigma)$, *formulas* $\varphi$ *of the language* $\mathcal{L}(\mathcal{S})$ *are given by*[4]

$$\varphi ::= X = x \mid \neg\varphi \mid \varphi \wedge \varphi \mid B^{\psi}\varphi \mid K^{\psi}\varphi \mid [\vec{V} = \vec{v}]\varphi$$

*where* $X \in \mathcal{U} \cup \mathcal{V}$, $x \in \Sigma$ *and* $\vec{V} = \vec{v}$ *is a sequence of the form* $V_1 = v_1, ..., V_n = v_n$ *where* $\vec{V} \in \mathcal{V}$.[5]

---

[4] $B\phi$ is seen as the abbreviation of $B^{\top}\phi$.

[5] For convenience, we will write both $V_1 = v_1, ..., V_n = v_n$ and $V_1 = v_1 \wedge ... \wedge V_n = v_n$ as $\vec{V} = \vec{v}$.

$\mathcal{L}(\mathcal{S})$ not only contains the doxastic operator $B$ (for belief) and $K$ (for knowledge) but also has the intervention operators of the form $[\vec{X} = \vec{x}]$ which expresses antecedents of counterfactuals. Therefore this language is able to express belief about counterfactuals and counterfactual beliefs $B[\vec{X} = \vec{x}]\phi$ and $[\vec{X} = \vec{x}]B\phi$.

For the semantics of $\mathcal{L}(\mathcal{S})$, we define the truth condition of counterfactual based on the causal epistemic model under the classical interventionist interpretation: a counterfactual $[\vec{X} = \vec{x}]\phi$ holds on a model $M$ whenever $\phi$ holds on the model $M_{\vec{X}=\vec{x}}$ which results from setting the value of $\vec{X}$ to $\vec{x}$. Therefore, we define the semantics of $\mathcal{L}(\mathcal{S})$ is given as below:

**Definition 5. (Semantics of the language $\mathcal{L}(\mathcal{S})$).**
*Let $M = \langle \mathcal{S}, \mathcal{F}, \leq, \mathcal{A} \rangle$ be a causal plausibility model.*

- *$\langle \mathcal{S}, \mathcal{F}, \leq, \mathcal{A} \rangle \models X = x$ iff $\mathcal{A}(X) = x$.*
- *$\langle \mathcal{S}, \mathcal{F}, \leq, \mathcal{A} \rangle \models B^\psi \phi$ iff $Min_\leq ||\psi|| \subseteq ||\phi||$, where $||\phi|| := \{\mathcal{A}' \in W^\mathcal{F} \mid \langle \mathcal{S}, \mathcal{F}, \leq, \mathcal{A}' \rangle \models \phi\}$.[6]*
- *$\langle \mathcal{S}, \mathcal{F}, \leq, \mathcal{A} \rangle \models K^\psi \phi$ iff $\langle \mathcal{S}, \mathcal{F}, \leq, \mathcal{A}' \rangle \models \phi$ for all $\mathcal{A}' \in ||\psi||$.*
- *$\langle \mathcal{S}, \mathcal{F}, \leq, \mathcal{A} \rangle \models [\vec{X} = \vec{x}]\phi$ iff $\langle \mathcal{S}, \mathcal{F}_{\vec{X}=\vec{x}}, \leq_{\vec{X}=\vec{x}} \mathcal{A}^\mathcal{F}_{\vec{X}=\vec{x}} \rangle \models \phi$, where $\leq_{\vec{X}=\vec{x}}$ is a total order over $W^{\mathcal{F}_{\vec{X}=\vec{x}}}$, defined as: $\mathcal{A}^\mathcal{F}_{\vec{X}=\vec{x}} \leq_{\vec{X}=\vec{x}} \mathcal{A}'^\mathcal{F}_{\vec{X}=\vec{x}}$ whenever $\mathcal{A}^\mathcal{F} \leq \mathcal{A}'^\mathcal{F}$.[7]*
- *the Boolean connectives are defined in the usual way.*

It is clear that the semantics is a combination of causal model and plausibility model. In particular, the intervention operator is treated as a typical dynamic operator in the style of dynamic epistemic logic developed extensively in the literature (see, e.g. [2,6,19]).

## 4.2   Important Notions that Are Expressible by $\mathcal{L}(\mathcal{S})$

In this section, we will introduce several important notions in the causality literature and show how to express them in our new language.

First, let us consider the concept of causal dependence. According to the classical definition of causal influence developed in [9], given a set of structural functions $\mathcal{F}$, an endogenous variable $Y$ *causally affects* $Z$ means there exist an assignment of some variables in $(\mathcal{U} \cup \mathcal{V}) \setminus \{Y, Z\}$, such that changing the value of $Y$ will force the value of $Z$ to be different[8]. The following proposition shows that the notion of causal influence is definable by $\mathcal{L}(\mathcal{S})$ as follows:

**Proposition 1.** *$Y$ causally affects $Z$ in $M$ iff*
$M \models \neg K \neg \bigvee_{\vec{X} \subseteq \mathcal{V} \setminus \{Y,Z\}, \vec{x}, y, z, z' \in \Sigma, z \neq z', Y \neq Z} ([\vec{X} = \vec{x}, Y = y]Z = z' \wedge [\vec{X} = \vec{x}]Z = z).$

---

[6] $Min_\leq S$ is defined as $\{w \in S \mid \forall t \in S, w \leq t\}$.

[7] $\leq_{\vec{X}=\vec{x}}$ is well-defined because $W^{\mathcal{F}_{\vec{X}=\vec{x}}}$ is identical to $\{\mathcal{A}^\mathcal{F}_{\vec{X}=\vec{x}} \mid \mathcal{A} \in W^\mathcal{F}\}$.

[8] Formally, $Y$ *causally affects* $Z$ in $M$ means there is an assignment $\mathcal{A}$ that complies with $\mathcal{F}$, a value $y \in \Sigma$, and a (partial) assignment to $\mathcal{V} \setminus \{Y, Z\}$ ($\vec{X} = \vec{x}$) such that $\mathcal{A}^\mathcal{F}_{\vec{X}=\vec{x},Y=y}(Y) \neq \mathcal{A}'^\mathcal{F}_{\vec{X}=\vec{x}}(Y)$.

*Proof.* $M \models \neg K \neg \bigvee_{\vec{X} \subseteq \mathcal{V} - \{Y,Z\}, \vec{x}, y, z, z' \in \Sigma, z \neq z', Y \neq Z}([\vec{X} = \vec{x}, Y = y]Z = z' \wedge [\vec{X} = \vec{x}]Z = z)$ iff there is some $\mathcal{A}' \in W^{\mathcal{F}}$ such that $\langle \mathcal{S}, \mathcal{F}, \leq, \mathcal{A}' \rangle \models \bigvee_{\vec{X} \subseteq \mathcal{V} - \{Y,Z\}, \vec{x}, y, z, z' \in \Sigma, z \neq z', Y \neq Z}$ $([\vec{X} = \vec{x}, Y = y]Z = z' \wedge [\vec{X} = \vec{x}]Z = z)$ iff there are some distinct variables $Y$ and $Z$, some $\vec{X} \subseteq \mathcal{V} - \{Y, Z\}$, such that $\mathcal{A}'^{\mathcal{F}}_{\vec{X} = \vec{x}, Y = y}(Z) \neq \mathcal{A}'^{\mathcal{F}}_{\vec{X} = \vec{x}}(Z)$. That is, at some possible state, forcing the value of some variable $Y$ to be $y$ changes the value of another variable $Z$, i.e., $Y$ causally affects $Z$. $\qquad\square$

Therefore we define the following abbreviation for causal influence, $Y \rightsquigarrow Z$
$$:= \neg K \neg \bigvee_{\vec{X} \subseteq \mathcal{V} \setminus \{Y,Z\}, \vec{x}, y, z, z' \in \Sigma, z \neq z', Y \neq Z}([\vec{X} = \vec{x}, Y = y]Z = z' \wedge [\vec{X} = \vec{x}]Z = z)$$

Next, the notion of *direct causal influence* can also be defined by our language. $Y$ has direct causal influence on $Z$, write $Y \rightsquigarrow^d Z$, is a special case of causal influence such that by fixing the value of *all* other variables, flip the value of $Y$, will change the value of $Z$. So we define the following abbreviation for direct causal influence:
$$Y \rightsquigarrow^d Z := \neg K \neg \bigvee_{\vec{X} = \mathcal{V} \setminus \{Y,Z\}, \vec{x}, y, z, z' \in \Sigma, z \neq z'}([\vec{X} = \vec{x}, Y = y]Z = z' \wedge [\vec{X} = \vec{x}]Z = z)$$

## 5  Doxastic Independence and Probabilistic Independence

### 5.1  Doxastic Independence

As we have seen, dependence and independence of belief play an important role in Example 1. Intuitively, the independence of belief between $\vec{X}$ and $\vec{Y}$ can be interpreted as the belief about $\vec{X}$ is invariant conditional on any (consistent) setting of $\vec{Y}$ and vice versa. With the language of doxastic causal reasoning, it can be formally expressed by:

$$Ind(\vec{X}, \vec{Y}) := \wedge_{\vec{x}, \vec{y} \in \Sigma}(\neg B^{\vec{Y} = \vec{y}} \perp \rightarrow (B^{\vec{Y} = \vec{y}}\vec{X} = \vec{x} \longleftrightarrow B\vec{X} = \vec{x}))$$
$$\vec{X} \perp\!\!\!\perp_B \vec{Y} := Ind(\vec{X}, \vec{Y}) \wedge Ind(\vec{Y}, \vec{X})$$

$\vec{X} \perp\!\!\!\perp_B \vec{Y}$ is the formal expression of *doxastic independence* between $\vec{X}$ and $\vec{Y}$ in our account. Similarly, we can define the *conditional doxastic independence* between $\vec{X}$ and $\vec{Y}$ given $\vec{Z}$ can be defined as:

$$Ind(\vec{X}, \vec{Y} \mid \vec{Z}) := \wedge_{\vec{x}, \vec{y}, \vec{z} \in \Sigma}(\neg B^{\vec{Y} = \vec{y}, \vec{Z} = \vec{z}} \perp \rightarrow (B^{\vec{Y} = \vec{y}, \vec{Z} = \vec{z}}\vec{X} = \vec{x} \longleftrightarrow B^{\vec{Z} = \vec{z}}\vec{X} = \vec{x}))$$
$$\vec{X} \perp\!\!\!\perp_B \vec{Y} \mid \vec{Z} := Ind(\vec{X}, \vec{Y} \mid \vec{Z}) \wedge Ind(\vec{Y}, \vec{X} \mid \vec{Z})$$

With these notions, we will show that our framework can explain very well the dependence and independence of belief in Example 1.

**Proposition 2.** *Let $\langle \mathcal{S}, \mathcal{F} \rangle$ be a causal model as shown in Fig. 1. Then:*

*(a) There is a uniform plausibility model $M$ of the form $\langle \mathcal{S}, \mathcal{F}, \leq, \mathcal{A} \rangle$ such that $M \not\models C \perp\!\!\!\perp_B P$*

*(b) for any $\leq$ and $\mathcal{A}$ such that $M = \langle \mathcal{S}, \mathcal{F}, \leq, \mathcal{A} \rangle$ is a uniform plausibility model, we have $M \models C \perp\!\!\!\perp_B P \mid T$*

*(c) There is a uniform plausibility model $M$ of the form $\langle \mathcal{S}, \mathcal{F}, \leq, \mathcal{A} \rangle$ such that $M \not\models C \perp\!\!\!\perp_B P \mid T, A$*

*Proof.* See appendix.

This result explains our intuition of dependence and independence in Example 1 in a precise manner. In addition, these results also fit the prediction made by the causal Bayesian network approach in a qualitative sense.

## 5.2    Comparison to Probabilistic Independence

The correspondence between belief and probability has been studied in [3], in which the Boolean value of $B(\vec{X} = \vec{x})$ (believing $\vec{X} = \vec{x}$) is seen as a qualitative representation of the probability of $\vec{X} = \vec{x}$ in a conditional probabilistic space. In probability theory, the probabilistic independence between $\vec{X}$ and $\vec{Y}$ (write $\vec{X} \perp\!\!\!\perp \vec{Y}$) with respect to a probabilistic distribution $P$ can be expressed as: for any value $\vec{x}$ of $\vec{X}$ and any value $\vec{y}$ of $\vec{Y}$, the conditional probability $P(\vec{X}=\vec{x} \mid \vec{Y}=\vec{y})$ is equal to the probability $P(\vec{X}=\vec{x})$. Following the correspondence between belief and probability argued in [3], we can interpret $B^{\vec{Y}=\vec{y}}\vec{X}=\vec{x} \leftrightarrow B\vec{X}=\vec{x}$ in the definition of $\perp\!\!\!\perp_B$ as a qualitative counterpart of $P(\vec{X}=\vec{x} \mid \vec{Y}=\vec{y}) = P(\vec{X}=\vec{x})$. Actually, our definition of doxastic independence preserves many important properties of probabilistic independence:

**Proposition 3.** *The following $\mathcal{L}(\mathcal{S})$-formulas are valid with respect to the class of causal plausibility models:*

a.  $(\vec{X} \perp\!\!\!\perp_B \vec{Y}|\vec{Z}) \to (\vec{Y} \perp\!\!\!\perp_B \vec{X}|\vec{Z})$ *(symmetry)*
b.  $(\vec{X} \perp\!\!\!\perp_B \vec{Y}\vec{W} \mid \vec{Z}) \to (\vec{X} \perp\!\!\!\perp_B \vec{Y} \mid \vec{Z})$ *(decomposition)*
c.  $(\vec{X} \perp\!\!\!\perp_B \vec{Y}\vec{W} \mid \vec{Z}) \to (\vec{X} \perp\!\!\!\perp_B \vec{Y} \mid \vec{Z}\vec{W})$ *(weak union)*
d.  $((\vec{X} \perp\!\!\!\perp_B \vec{Y} \mid \vec{Z}) \wedge (\vec{X} \perp\!\!\!\perp_B \vec{W} \mid \vec{Z}\vec{Y})) \to (\vec{X} \perp\!\!\!\perp_B \vec{Y}\vec{W} \mid \vec{Z})$ *(contraction)*
e.  $((\vec{X} \perp\!\!\!\perp_B \vec{W} \mid \vec{Z}\vec{Y}) \wedge (\vec{X} \perp\!\!\!\perp_B \vec{Y} \mid \vec{Z}\vec{W})) \to (\vec{X} \perp\!\!\!\perp_B \vec{Y}\vec{W} \mid \vec{Z})$ *(intersection)*

*Proof.* See appendix.

## 5.3    Relation with Causal Bayesian Network

It is well-known that some quantitative modelling approaches (such as causal Bayesian networks) are very successful in characterizing dependence and independence in a causal structure. For instance, from the perspective of a causal Bayesian network, Example 1 can be formalized as the directed acyclic graph in Fig. 1. Based on this graph, (i) and (ii) can be justified by the notion of "d-separation".[9]

According to [20], $\vec{X}$ and $\vec{Y}$ are independent conditional on $\vec{Z}$ for any probabilistic distribution (which is Markovian relative to the causal graph) whenever $\vec{X}$ and $\vec{Y}$ are d-separated by $\vec{Z}$. Thus:

---

[9] Path is a notion of directed acyclic graph which means a sequence of arrows in the graph. We use "$\twoheadrightarrow$" and "$\twoheadleftarrow$" to denote arrows in the graph. A path $p$ is d-separated by a set of variables $Z$ iff (i) $p$ contains $i \twoheadrightarrow m \twoheadrightarrow j$ or $i \twoheadleftarrow m \twoheadrightarrow j$ such that $m \in Z$, or (ii) $p$ contains $i \twoheadrightarrow m \twoheadleftarrow j$ such that $m \notin Z$ and no descendant of $m$ is in $Z$. Given two sets of variables $\vec{X}$ and $\vec{Y}$, $\vec{X}$ and $\vec{Y}$ are d-separated by $\vec{Z}$ if and only if $\vec{Z}$ d-separate every path from $\vec{X}$ to $\vec{Y}$.

(a') The independence between $C$ and $P$ is not guaranteed as they are **not** d-separated by $\varnothing$

(b') $C$ and $P$ are independent conditional on $T$ as they are d-separated by $\{T\}$.

(c') The independence between $C$ and $P$ conditional on $A, T$ is not guaranteed as they are **not** d-separated by $\{A, T\}$

We can find that (a),(b),(c) corresponds to (a'),(b'),(c') in Proposition 2. Though they are derived from two different characterizations of the causal structure, there is a clear correspondence between the quantitative approach and the qualitative approach.

## 6    Axiomatization

The logic of doxastic causal reasoning can be axiomatized by the Hilbert style system $\mathsf{L}_{BCU}$, whose axioms and rules are given in Table 1. Since the axioms of $\mathsf{L}_{BCU}$ depend on the signature of the language, we will write the axiom system for the language $\mathcal{L}(S)$ as $\mathsf{L}_{BCU}(S)$.

Our axiomatization is both based on the axiom system of counterfactuals developed in [9] and the axiom system of conditional doxastic logic developed in [1,4,7]. However, we are not just simply merging the axioms from the two logic systems. There are interesting new axioms in $\mathsf{L}_{BCU}$ which characterize the interaction between knowledge, belief, and causality. Those axioms reflect the typical features of doxastic causal reasoning, and they can not be derived from the logic of counterfactuals and logic of conditional belief.

The axioms of the system $\mathsf{L}_{BCU}$ can be sorted into three kinds.

The first kind of the axioms, which includes $A_1$ to $A_5$, $A_\neg$, $A_\wedge$ and $A_{[]}$, is directly from the system $AX_{rec}$ in [9] or developed in [10]. They describe how the intervention operator works in the causal structure. $A_1$ to $A_4$ express the functionality of intervention. $A_5$ guarantees the causal influence is acyclic. $A_\neg$, $A_\wedge$ and $A_{[]}$ are reduction axioms for Boolean operators.

The second kind of axioms includes $B1$ to $B6$. Those axioms are from the axiom system developed in [1,4,7] and they characterize the properties of conditional belief and knowledge.

The third kind of axioms includes $KB$, $CM$, $SD$, $IGN$ and $UNI$. Those axioms characterize the entanglement between the causal structure and epistemic operators. Axiom $KB$ is a reduction axiom for knowledge. Axiom $CM$ expresses that intervention does not add or reduce the information of an agent (an agent believes $\phi$ after an intervention whenever the agent believes $\phi$ holds after the intervention). Axiom $SD$ expresses that the agent has full knowledge of causality and the world is semi-deterministic: if $\phi$ is possible, then had the agent hypothetically got all of the information about the exogenous variables, $\phi$ would be certain to the agent. Axiom $IGN$ expresses that the agent is ignorant about the real value of exogenous variables. Axiom $UNI$ expresses that all exogenous variables are independent of each other.

**Table 1.** Axiom System $\mathsf{L}_{BCU}$

| P | $\phi$ for $\phi$ an instance of a propositional tautology | MP | From $\varphi_1$ and $\varphi_1 \to \varphi_2$ infer $\varphi_2$ |
|---|---|---|---|
| Nec | From $\varphi$ infer $[\vec{X} = \vec{x}]\varphi$ and $B^{\psi}\varphi$ | LE | From $\varphi \leftrightarrow \psi$ infer $B^{\phi}\chi \leftrightarrow B^{\psi}\chi$ |
| $A_1$ | $[\vec{X}=\vec{x}]Y=y \to \neg[\vec{X}=\vec{x}]Y=y'$ for $y, y' \in \{0, 1\}$ with $y \neq y'$ | $A_2$ | $\bigvee_{y\in\Sigma}[\vec{X}=\vec{x}]Y=y$ |
| $A_3$ | $([\vec{X}=\vec{x}]Y=y \wedge [\vec{X}=\vec{x}]Z=z) \to [\vec{X}=\vec{x}, \vec{Y}=\vec{y}]Z=z$ | $A_4$ | $[\vec{X}=\vec{x}, Y=y]Y=y$ |
| $A_5$ | $(X_0 \rightsquigarrow X_1 \wedge \cdots \wedge X_{k-1} \rightsquigarrow X_k) \to \neg(X_k \rightsquigarrow X_0)$ | | |
| | $X_0, \ldots, X_k$ are distinct variables in $\mathcal{V}$ | | |
| $A_\neg$ | $[\vec{X}=\vec{x}] \neg \varphi \leftrightarrow \neg[\vec{X}=\vec{x}]\varphi$ | $A_\wedge$ | $[\vec{X} = \vec{x}]\varphi_1 \wedge \varphi_2 \leftrightarrow [\vec{X} = \vec{x}]\varphi_1 \wedge [\vec{X} = \vec{x}]\varphi_2)$ |
| $A_{[][]}$ | $[\vec{X} = \vec{x}][\vec{Y} = \vec{y}]\varphi \leftrightarrow [\vec{X} = \vec{x}', \vec{Y} = \vec{y}]\phi$ | | |
| KB | $K^{\psi}\phi \leftrightarrow B^{\psi \wedge \neg\phi}\phi$ | CM | $[\vec{X} = \vec{x}]B^{\psi}\phi \leftrightarrow B^{\psi}[\vec{X} = \vec{x}]\phi$ |
| SD | $\neg K \neg (\vec{U} = \vec{u} \wedge \phi) \to K^{\vec{U}=\vec{u}}\phi$ if $\{\vec{U}\} = \mathcal{U}$ | IGN | $\neg K \neg (\vec{U} = \vec{u})$ if $\vec{U} \in \mathcal{U}$ |
| UNI | $\vec{U} \perp\!\!\!\perp_B \vec{U}' \mid \vec{U}''$ for disjoint $\vec{U}, \vec{U}'$ and $\vec{U}''$ in $\mathcal{U}$ | | |
| B1 | $B^{\chi}(\phi \to \psi) \to (B^{\chi}\phi \to B^{\chi}\psi)$ | B2 | $K\phi \to \phi$ |
| B3 | $K\phi \to B^{\psi}\phi$ | B4 | $B^{\chi}\phi \to KB^{\chi}\phi \qquad \neg B^{\chi}\phi \to K \neg B^{\chi}\phi$ |
| B5 | $B^{\phi}\phi$ | B6 | $\neg B^{\phi} \neg \psi \to (B^{\phi \wedge \psi}\chi \leftrightarrow B^{\phi}(\psi \to \chi))$ |

The deduction rule of $\mathsf{L}_{BCU}$ includes the $MP$ rule, Necessitation rule, and LE rule as conditional doxastic logic.

Actually the Axiom $UNI$ defines the property of being uniform:

**Proposition 4.** *A basic causal plausibility model $\langle \mathcal{S}, \mathcal{F}, \leq, \mathcal{A} \rangle$ is uniform iff for any disjoint sequences of exogenous variable $\vec{U}, \vec{U}'$ and $\vec{U}''$, $\langle \mathcal{S}, \mathcal{F}, \leq, \mathcal{A} \rangle \vDash \vec{U} \perp\!\!\!\perp_B \vec{U}' \mid \vec{U}''$.*

*Proof.* See appendix.

Let $\mathsf{L}_{BC}$ be the fragment of $\mathsf{L}_{BCU}$ which excludes the axiom $IGN$ from $\mathsf{L}_{BCU}$. We can prove the following completeness theorem for the logic of doxastic causal reasoning:

**Theorem 1.** *(Completeness theorem for $\mathsf{L}_{BC}$ and $\mathsf{L}_{BCU}$)*

(a) *$\mathsf{L}_{BC}(\mathcal{S})$ is sound and strongly complete for $\mathcal{L}(\mathcal{S})$ with respect to the class of all basic causal plausibility models.*

(b) *$\mathsf{L}_{BCU}(\mathcal{S})$ is sound and strongly complete for $\mathcal{L}(\mathcal{S})$ with respect to the class of all uniform causal plausibility models.*

*Proof.* **(a):** The proof of soundness can be seen in appendix. For the completeness of $\mathsf{L}_{BC}$, it is sufficient to show that for any maximal consistent set of $\mathcal{L}(\mathcal{S})$-formulas of $\mathsf{L}_{BC}$, there is a causal plausibility model.

Before proceeding, let us define some fragments of $\mathcal{L}(\mathcal{S})$. Let $\mathcal{L}^-(\mathcal{S})$ be the fragment of $\mathcal{L}(\mathcal{S})$ in which there is not epistemic operator nested in intervention operators and $\mathcal{L}^0(\mathcal{S})$ be the fragment of $\mathcal{L}^-(\mathcal{S})$ which excludes epistemic operators. By the axiom $KB, CM, A_\wedge, A_\neg$, every formula $\phi$ of $\mathcal{L}(\mathcal{S})$ can be reduced to a formula $tr(\phi) \in \mathcal{L}^-(\mathcal{S})$ such that $\vdash_{BC} \phi \leftrightarrow tr(\phi)$. Therefore, it is sufficient to show that for any maximal consistent set of $\mathcal{L}^-(\mathcal{S})$-formulas $\Gamma$ of $\mathsf{L}_{BC}$, there is a basic causal plausibility model $\langle \mathcal{S}, \mathcal{F}^\Gamma, \leq^\Gamma, \mathcal{A}^\Gamma \rangle \vDash \Gamma$

$\mathcal{A}^\Gamma$ is defined in the obvious way, that is $\mathcal{A}^\Gamma(X)=x$ whenever $X=x \in \Gamma$. $A_1$ and $A_2$ guarantee that it is well defined. $\mathcal{F}^\Gamma=\{f_V \mid V \in \mathcal{V}\}$ is defined as follows: for each $V \in \mathcal{V}$, $f_V^\Gamma$ is a function such that $f_V^\Gamma(\vec{U}=\vec{u}, \vec{X}=\vec{x})=v$ iff $K^{\vec{U}=\vec{u}}[\vec{X}=\vec{x}]V=v \in \Gamma$ where $\vec{U}$ are all the exogenous variables and $\vec{X}$ are all the endogenous variables in $\mathcal{V}\setminus\{V\}$. $A_1$, $A_2$ and $SD$ guarantees that there is a unique $\sigma \in \Sigma$ such that $K^{\vec{U}=\vec{u}}[\vec{X}=\vec{x}]V=\sigma \in \Gamma$. Therefore $f_V$ is well defined. In particular, we have:

**Lemma 1.** *If $\vec{U}=\vec{u} \in \Gamma$, then $f_V^\Gamma(\vec{U}=\vec{u}, \vec{X}=\vec{x})=v$ iff $[\vec{X}=\vec{x}]V=v \in \Gamma$. (The proof of Lemma 1 can be seen in appendix.)*

By the same argument as [9], we can conclude that:

$$\text{If } \chi \in \mathcal{L}^0(\mathcal{S}), \text{ then for any } \leq, \langle \mathcal{S}, \mathcal{F}^\Gamma, \leq, \mathcal{A}^\Gamma \rangle \vDash \chi \text{ iff } \chi \in \Gamma. \ (*)$$

Then we will construct the plausibility ordering $\leq^\Gamma$. We can think of the formulas in $\mathcal{L}^0(\mathcal{S})$ as atomic propositional symbols, and follow the construction of canonical models for the $BRSI$ system in [7]. We first define a series of orderings $\preccurlyeq^w$ on $\mathsf{L}_{BC}$-consistent sets of $\mathcal{L}^-(\mathcal{S})$ formulas (where $w$ is a $\mathsf{L}_{BC}$-consistent set). For maximal consistent sets of $\mathcal{L}^-(\mathcal{S})$-formulas $w, t, u$: we define $t \preccurlyeq^w u$ whenever there is some $\phi \in t \cap u$ such that $\{\psi \mid B^\phi \psi \in w\} \subseteq T$. Let $W^w=\{x \mid x \preccurlyeq^w y \text{ for some } y\}$. Following exactly the same steps as in [7], it can be shown that:

**Lemma 2.** $B^\phi \psi \in w$ iff $Min_{\preccurlyeq^\Gamma} |\phi|_w \subseteq |\psi|_w$ where $|\phi|_w$ refers to $\{s \in W^w \mid \phi \in s\}$.

In addition, [7] shows that by Axiom $B4$, for any $s \in W^w$, $\preccurlyeq^s$ and $\preccurlyeq^w$ are identical.

**Lemma 3.** *for each assignment to all exogenous variables $\vec{U}=\vec{u}$, there is exactly one $\Theta \in W^\Gamma$ such that $\vec{U}=\vec{u} \in \Theta$. (The proof of Lemma 3 can be seen in appendix.)*

By definition, for each full assignment to exogenous variables $\vec{U}=\vec{u}$, there is exactly one assignment $\mathcal{A} \in W^\Gamma$ with $\mathcal{A}(\vec{U})=\vec{u}$. So there is one-to-one correspondence between the members of $W^\Gamma$ and the assignments in $W^\mathcal{F}$ (the set of all assignments that complies with $\mathcal{F}^\Gamma$). By this bijection, we can define $\leq^\Gamma$ as follows: For any assignments $\mathcal{A}_1, \mathcal{A}_2 \in W^\mathcal{F}$, $\mathcal{A}_1 \leq^\Gamma \mathcal{A}_2$ iff $w_1 \preccurlyeq^\Gamma w_2$ where $w_n$ refers to the unique assignment in $W^\Gamma$ with $\mathcal{A}_n=\mathcal{A}^{w_n}$. Therefore $(W^\mathcal{F}, \leq^\Gamma)$ is isomorphic to $(W^\Gamma, \preccurlyeq^\Gamma)$.

Then we are ready to show that for any maximal consistent set of $\mathcal{L}^-(\mathcal{S})$-formulas $\Gamma$ of $\mathsf{L}_{BC}$, $\phi \in \Gamma \ \langle \mathcal{S}, \mathcal{F}^\Gamma, \leq^\Gamma, \mathcal{A}^\Gamma \rangle \vDash \phi$. By induction on the complexity of $\phi$, we have:

- If $\chi \in \mathcal{L}^0(\mathcal{S})$, then by (*), $\phi \in \Gamma \ \langle \mathcal{S}, \mathcal{F}^\Gamma, \leq^\Gamma, \mathcal{A}^\Gamma \rangle \vDash \chi$. The Boolean cases are trivial.
- If $\chi$ is of the form $B_\psi^\phi$ $B^\phi \psi \in \Gamma$ iff $B^\phi \psi \in \Gamma$ iff $Min_{\preccurlyeq^\Gamma} |\phi|_\Gamma \subseteq |\psi|_\Gamma$ (by Lemma 2) iff $Min_{\leq} ||\phi|| \subseteq ||\psi||$ where $||\phi|| := \{\mathcal{A}' \in W^\mathcal{F} \mid \langle \mathcal{S}, \mathcal{F}^\Gamma, \leq^\Gamma, \mathcal{A}' \rangle \vDash \phi\}$ (by the inductive hypothesis and the isomorphism between $\langle W^\mathcal{F}, \leq^\Gamma \rangle$ and $\langle W^\Gamma, \preccurlyeq^\Gamma \rangle$) iff $B^\phi \psi \in \Gamma$.

In addition, based on this result, $A5$ guarantees that (by Proposition 1) $\mathcal{F}^{\Gamma}$ is acyclic. So $\langle \mathcal{S}, \mathcal{F}^{\Gamma}, \leq^{\Gamma}, \mathcal{A}^{\Gamma} \rangle$ fulfills all the requirement of basic causal plausibility models.

**(b):** By the completeness result of Theorem 1(a), for any maximal $\mathsf{L}_{BC}$-consistent set $\Gamma$, there is $\langle \mathcal{S}, \mathcal{F}^{\Gamma}, \leq^{\Gamma}, \mathcal{A}^{\Gamma} \rangle \vDash \Gamma$. If $\Gamma$ is $\mathsf{L}_{BCU}$ consistent, then for any disjoint sequences of exogenous variable $\vec{U}$, $\vec{U}'$ and $\vec{U}''$, $\vec{U} \perp\!\!\!\perp_B \vec{U}' \mid \vec{U}'' \in \Gamma$. By Proposition 4, $\langle \mathcal{S}, \mathcal{F}^{\Gamma}, \leq^{\Gamma}, \mathcal{A}^{\Gamma} \rangle$ must be uniform. So every $\mathsf{L}_{BCU}$-consistent set $\Gamma$ has a uniform causal plausibility model.By Proposition 4, Axiom $UNI$ is sound with respect to the class of uniform causal plausibility models. $\qquad\square$

To define the notion of causal dependence, we assume that the signature $\mathcal{S}$ is finite. The assumption makes the decidability problem trivial: if $\mathcal{S}$ is finite, there are only finitely many causal models based on $\mathcal{S}$. However, we can show that the problem of satisfiability is still decidable even with an infinite signature.

**Proposition 5.** *A $\mathcal{L}(\mathcal{S})$ formula $\phi$ is satisfiable in a model based on a signature $\mathcal{S}$ iff it is satisfiable in a model based on a finite signature.*

*Proof.* See appendix.

# 7    Conclusion and Future Work

In this paper, we have proposed an account of doxastic causal reasoning based on integrating the existing causal and plausibility models. Our formal framework includes both the traditional interventionist causal language and epistemic operators for belief revision so that it is able to express important concepts and characterize the reasoning of causality and dynamic change of beliefs. Technically, we developed a complete deductive system for doxastic causal reasoning, and its satisfiability problem is decidable. In addition, we illustrated with examples that our qualitative approach makes the same prediction of dependence/independence as the quantitative account in terms of the causal Bayesian network. For future directions, we plan to investigate further issues concerning the contrast between qualitative and quantitative approaches. Also, we want to extend the formal framework to a multi-agent setting so that our account can be used to model how a group of agents does causal reasoning when each of the agents gets different information.

**Acknowledgements.** Many thanks to Johan van Benthem, Kevin Kelly and Zhiguang Zhao for their suggestions. We benefited greatly from discussions with other members of Tsinghua University-University of Amsterdam Joint Research Centre for Logic. Thanks also for the three anonymous reviewers for their insightful comments for improvements.

# A   Appendix

**Proof of Proposition 2**

Let assignments as shown in the following table. Let the ordering $\leq$ be: $\mathcal{A}_1 \leq \mathcal{A}_2 \leq \cdots \leq \mathcal{A}_{16}$. It is easy to see that the ordering is uniform. And we can see that $\langle \mathcal{S}, \mathcal{F}, \leq, \mathcal{A} \rangle \vDash \neg B^{P=1} \perp$ but $\langle \mathcal{S}, \mathcal{F}, \leq, \mathcal{A} \rangle \nvDash B^{C=1}P=1 \leftrightarrow BP=1$. $\square$

| | $U_C$ | $U_P$ | $T$ | $U_A$ | $C$ | $P$ | $A$ | | $U_C$ | $U_P$ | $T$ | $U_A$ | $C$ | $P$ | $A$ | | $U_C$ | $U_P$ | $T$ | $U_A$ | $C$ | $P$ | $A$ | | $U_C$ | $U_P$ | $T$ | $U_A$ | $C$ | $P$ | $A$ |
|---|---|---|---|---|---|---|---|---|---|---|---|---|---|---|---|---|---|---|---|---|---|---|---|---|---|---|---|---|---|---|---|
| $\mathcal{A}_1$ | 1 | 1 | 0 | 1 | 0 | 0 | 0 | $\mathcal{A}_5$ | 1 | 1 | 1 | 1 | 1 | 1 | 1 | $\mathcal{A}_9$ | 1 | 1 | 0 | 0 | 0 | 0 | 0 | $\mathcal{A}_{13}$ | 1 | 1 | 1 | 0 | 1 | 1 | 0 |
| $\mathcal{A}_2$ | 0 | 1 | 0 | 1 | 0 | 0 | 0 | $\mathcal{A}_6$ | 0 | 1 | 1 | 1 | 0 | 1 | 1 | $\mathcal{A}_{10}$ | 0 | 1 | 0 | 0 | 0 | 0 | 0 | $\mathcal{A}_{14}$ | 0 | 1 | 1 | 0 | 0 | 1 | 0 |
| $\mathcal{A}_3$ | 1 | 0 | 0 | 1 | 0 | 0 | 0 | $\mathcal{A}_7$ | 1 | 0 | 1 | 1 | 1 | 0 | 1 | $\mathcal{A}_{11}$ | 1 | 0 | 0 | 0 | 0 | 0 | 0 | $\mathcal{A}_{15}$ | 1 | 0 | 1 | 0 | 1 | 0 | 0 |
| $\mathcal{A}_4$ | 0 | 0 | 0 | 1 | 0 | 0 | 0 | $\mathcal{A}_8$ | 0 | 0 | 1 | 1 | 1 | 0 | 0 | $\mathcal{A}_{12}$ | 0 | 0 | 0 | 0 | 0 | 0 | 0 | $\mathcal{A}_{16}$ | 0 | 0 | 1 | 0 | 0 | 0 | 0 |

(b) is easy to check by the semantics.

(c): By the table in (a), we can see that $\langle \mathcal{S}, \mathcal{F}, \leq, \mathcal{A} \rangle \vDash \neg B^{T=1,A=0} \perp$ but $\langle \mathcal{S}, \mathcal{F}, \leq, \mathcal{A} \rangle \nvDash B^{T=1,A=0,C=1}P=1 \leftrightarrow B^{T=1,A=0}P=1$.

**Proof of Proposition 3** (a) is obtained by the definition of conditional independence directly. For the convenience of the following proof, let $M$ be any uniform causal plausibility model, we show that $M \vDash Ind(\vec{X}, \vec{Y} \mid \vec{Z}) \rightarrow Ind(\vec{Y}, \vec{X} \mid \vec{Z})$. Suppose that $M \vDash Ind(\vec{X}, \vec{Y} \mid \vec{Z})$. If there exist $\vec{z}, \vec{x}, \vec{y} \in \Sigma$ such that $Min_{\leq}||\vec{Z}=\vec{z} \wedge \vec{X}=\vec{x}|| \subseteq ||\vec{Y}=\vec{y}||$ and $Min_{\leq}||\vec{Z}=\vec{z}|| \nsubseteq ||\vec{Y}=\vec{y}||$, then there exist $\mathcal{A}_1 \in Min_{\leq}||\vec{Z}=\vec{z}||$ with $\mathcal{A}_1(\vec{X})=\vec{x_1} \neq \vec{y}$ and $\mathcal{A}_1(\vec{Y})=\vec{y_1} \neq \vec{y}$, $\mathcal{A}_2 \in Min_{\leq}||\vec{X}=\vec{x} \wedge \vec{Z}=\vec{z}||$ with $\mathcal{A}_2(\vec{Y})=\vec{y}$. Since causal models are acyclic, there are two cases: (1) the values of $\vec{Y}$ are determined by the values of $\vec{Z}$. Then $\mathcal{A}_1(\vec{Y}) = \mathcal{A}_2(\vec{Y})$ since $\mathcal{A}_1(\vec{Z}) = \mathcal{A}_2(\vec{Z})$, a contradiction. (2): there are disjoint sequences of exogenous variables $\vec{U}$ and $\vec{U'}$ such that $\vec{U}$ determines the values of $\vec{X}\vec{Z}$. and $\vec{U'}$ determines the values of $\vec{Y}$. Hence, there exist $\vec{u_1}, \vec{u_2}, \vec{u'_1}, \vec{u'_2} \in \Sigma$ such that $\mathcal{A}_1(\vec{U})=\vec{u_1}$, $\mathcal{A}_1(\vec{U'})=\vec{u'_1}$ and $\mathcal{A}_2(\vec{U})=\vec{u_2}$, $\mathcal{A}_2(\vec{U'})=\vec{u'_2}$. Then there exist $\mathcal{A}_3$ and $\mathcal{A}_4$ where $\mathcal{A}_3(\vec{U})=\vec{u_1}$, $\mathcal{A}_3(\vec{U'})=\vec{u'_2}$, $\mathcal{A}_3(\vec{X})=\vec{x_1}$, $\mathcal{A}_3(\vec{Y})=\vec{y}$ and $\mathcal{A}_4(\vec{U})=\vec{u_2}$, $\mathcal{A}_4(\vec{U'})=\vec{u'_1}$, $\mathcal{A}_4(\vec{X})=\vec{x}$, $\mathcal{A}_3(\vec{Y})=\vec{y_1}$, $\mathcal{A}_3(\vec{Z})=\vec{z}=\mathcal{A}_4(\vec{Z})$. Since $M$ is uniform, $\mathcal{A}_1 \leq \mathcal{A}_3$ implies $\mathcal{A}_4 \leq \mathcal{A}_2$. Since $\mathcal{A}_1 \in Min_{\leq}||\vec{Z}=\vec{z}||$ and $\mathcal{A}_2 \in Min_{\leq}||\vec{Z}=\vec{z} \wedge \vec{X}=\vec{x}||$, we have $\mathcal{A}_1 \leq \mathcal{A}_3$ and $\mathcal{A}_2 \leq \mathcal{A}_4$, a contradiction. The other direction is similar. Hence, we have $M \vDash Ind(\vec{Y}, \vec{X} \mid \vec{Z})$.

The proofs of (b)(c)(d)(e) are similar. We show the proof of (b) as an example: Suppose that $M \vDash (\vec{X} \perp\!\!\!\perp_B \vec{Y}\vec{W} \mid \vec{Z})$. For any $\vec{z}, \vec{y}, \vec{x} \in \Sigma$: Assume that $Min_{\leq}||\vec{Z}=\vec{z} \wedge \vec{X}=\vec{x}|| \subseteq ||\vec{Y}=\vec{y}||$. Since $\leq$ is a total order, there exists $\vec{w'} \in \Sigma$ such that $Min_{\leq}||\vec{Z}=\vec{z} \wedge \vec{X}=\vec{x}||=Min_{\leq}||\vec{Z}=\vec{z} \wedge \vec{X}=\vec{x} \wedge \vec{W}=\vec{w'}||$. Then $Min_{\leq}||\vec{Z}=\vec{z} \wedge \vec{X}=\vec{x}|| \subseteq ||\vec{Y}=\vec{y} \wedge \vec{W}=\vec{w'}||$. It follows that $Min_{\leq}||\vec{Z}=\vec{z}|| \subseteq ||\vec{Y}=\vec{y} \wedge \vec{W}=\vec{w'}|| \subseteq ||\vec{Y}=\vec{y}||$. Assume that $Min_{\leq}||\vec{Z}=\vec{z}|| \subseteq ||\vec{Y}=\vec{y}||$. Similarly, there exists $\vec{w'} \in \Sigma$ such that $Min_{\leq}||\vec{Z}=\vec{z}|| \subseteq ||\vec{Y}=\vec{y} \wedge \vec{W}=\vec{w'}||$. Hence, $Min_{\leq}||\vec{Z}=\vec{z} \wedge \vec{X}=\vec{x}|| \subseteq ||\vec{Y}=\vec{y} \wedge \vec{W}=\vec{w'}|| \subseteq ||\vec{Y}=\vec{y}||$. So we have $M \vDash Ind(\vec{X}, \vec{Y} \mid \vec{Z})$. $\square$

**Proof of Proposition** 4 Let $M=\langle\mathcal{S},\mathcal{F},\leq,\mathcal{A}\rangle$ be any causal plausibility model.

$\Rightarrow$: Suppose that there exist $\vec{U},\vec{U}'$ and $\vec{U}''$ such that $M\not\models\vec{U}\perp\!\!\!\perp_B\vec{U}'\mid\vec{U}''$. We consider the case that $M\models B^{\vec{U}=\vec{u_0},\vec{U}''=\vec{u}''}\vec{U}'=\vec{u_0}'$ and $M\not\models B^{\vec{U}''=\vec{u}''}\vec{U}'=\vec{u_0}'$. Then there exists $\mathcal{A}_1\in Min_\leq||\vec{U}''=\vec{u}''||$, such that $\mathcal{A}_1(\vec{U}')=\vec{u_1}'\neq\vec{u_0}'$ and $\mathcal{A}_1(\vec{U})=\vec{u_1}\neq\vec{u_0}$. Let $\mathcal{A}_2,\mathcal{A}_3,\mathcal{A}_4$ be assignments such that $\mathcal{A}_2(\vec{U}')=\vec{u_0}'$, $\mathcal{A}_3(\vec{U})=\vec{u_0}$, $\mathcal{A}_4(\vec{U}')=\vec{u_1}'$ and $\mathcal{A}_2,\mathcal{A}_3,\mathcal{A}_4$ agree the values of the other exogenous variables with $\mathcal{A}_1,\mathcal{A}_2,\mathcal{A}_3$ respectively. Let $\vec{U}^-=\mathcal{U}-\{\vec{U}'\}$. Then $\mathcal{A}_1(\vec{U}^-)=\mathcal{A}_2(\vec{U}^-)$ and $\mathcal{A}_3(\vec{U}^-)=\mathcal{A}_4(\vec{U}^-)$, $\mathcal{A}_1(\vec{U}')=\mathcal{A}_4(\vec{U}')$ and $\mathcal{A}_2(\vec{U}')=\mathcal{A}_3(\vec{U}')$ but $\mathcal{A}_4\not\leq\mathcal{A}_3$. Hence, $M$ is not uniform.

$\Leftarrow$: For any assignments $\mathcal{A}_1,\mathcal{A}_2$ and $\vec{U}\in\mathcal{U}$, let $\vec{U}^-=\mathcal{U}-\{\vec{U}\}$. Suppose that $\mathcal{A}_1\leq\mathcal{A}_2$ and $\mathcal{A}_1(\vec{U}^-)=\mathcal{A}_2(\vec{U}^-)=\vec{u_1}^-$. Let $\mathcal{A}_1'$ and $\mathcal{A}_2'$ be assignments such that $\mathcal{A}_1'(\vec{U})=\mathcal{A}_1(\vec{U})=\vec{u_1},\mathcal{A}_2'(\vec{U})=\mathcal{A}_2(\vec{U})=\vec{u_2}$ and $\mathcal{A}_1'(\vec{U}^-)=\mathcal{A}_2'(\vec{U}^-)=\vec{u_2}^-$. Since $\leq$ is a total order and $\mathcal{A}_1\leq\mathcal{A}_2$, there exists $\vec{U}^*=\vec{u}^*$ where $\vec{U}^*$ is a sequence of exogenous variables and $\vec{u}^*\in\Sigma$ such that $\mathcal{A}_1\in Min_\leq||\vec{U}^*=\vec{u}^*||$. If $\vec{U}^*\subseteq\vec{U}$ or $\vec{U}^*\subseteq\vec{U}^-$, then we have $\mathcal{A}_1'\leq\mathcal{A}_2'$ since $\vec{U}$ is independent from $\vec{U}^-$. If $\vec{U}^*\subseteq\vec{U}\cup\vec{U}^*$, then there exists $\vec{U}'\subseteq\vec{U}$ and $\vec{U}''\subseteq\vec{U}^-$ such that $\vec{U}'\cup\vec{U}''=\vec{U}^*$. Let $\vec{U}_1=\vec{U}-\vec{U}'$ and $\vec{U}_2=\vec{U}^--\vec{U}''$. Note that $\vec{U}_1,\vec{U}^*$ and $\vec{U}_2$ are disjoint, and $\vec{U}_1',\vec{U}_2',\vec{U}^*$ are independent from each other. Let $\mathcal{A}_1(\vec{U}_1')=\vec{u}_1'=\mathcal{A}_1'(\vec{U}_1'),\mathcal{A}_1(\vec{U}_2')=\vec{u}_2'=\mathcal{A}_2(\vec{U}_2')$, $\mathcal{A}_1(\vec{U}^*)=\vec{u}^*$ and $\mathcal{A}_1'(\vec{U}_2')=\vec{u}_2''=\mathcal{A}_2'(\vec{U}_2')$. Then $M\models B^{\vec{U}_2=\vec{u}_2',\vec{U}^*=\vec{u}^*}\vec{U}_1=\vec{u}_1'$ and $M\models B^{\vec{U}_2=\vec{u}_2'}\vec{U}_1=\vec{u}_1'$. We have $\mathcal{A}_1'\in Min_\leq||\vec{U}_2=\vec{u}_2''||\subseteq||\vec{U}_1=\vec{U}_1'||$. Since $\mathcal{A}_2'\in||\vec{U}_2=\vec{u}_2''||$, we have $\mathcal{A}_1'\leq\mathcal{A}_2'$. So $M$ is uniform.  $\square$

**Proof of Theorem** 1(a)**Soundness** Let $M=\langle\mathcal{S},\mathcal{F},\leq,\mathcal{A}\rangle$ be a causal plausibility model.

For Axiom $KB$, $Min_\leq||\neg\phi||\subseteq||\phi||$ iff $||\neg\phi||=\varnothing$ iff $K\phi$ holds.

By Definition 4.1, $\mathcal{A}\leq\mathcal{A}'\Leftrightarrow\mathcal{A}^{\mathcal{F}}_{\vec{X}=\vec{x}}\leq\mathcal{A}'^{\mathcal{F}}_{\vec{X}=\vec{x}}$. Therefore $Min_\leq||[\vec{X}=\vec{x}]\psi||\subseteq||[\vec{X}=\vec{x}]\phi||$ iff $Min_{\leq_{\vec{X}=\vec{x}}}||\psi||\subseteq||\phi||$. So Axiom $CM$ is sound.

If $M\models\neg K\neg(\vec{U}=\vec{u}\wedge\phi)$, then there is $\mathcal{A}'\in W^{\mathcal{F}}$ such that $\langle\mathcal{S},\mathcal{F},\leq,\mathcal{A}'\rangle\models\vec{U}=\vec{u}\wedge\phi$. Since $\mathcal{F}$ is acyclic, and $\vec{U}$ are all the exogenous variables, there is exactly one $\mathcal{A}'\in W^{\mathcal{F}}$ such that $\mathcal{A}'(\vec{U})=\vec{u}$. So $||\vec{U}=\vec{u}||=\{\mathcal{A}'\}$ and $\langle\mathcal{S},\mathcal{F},\leq,\mathcal{A}'\rangle\models\phi$. Therefore $M\models K^{\vec{U}=\vec{u}}\phi$. So Axiom $SD$ is sound.

As for each $\vec{U}=\vec{u}$, there is $\mathcal{A}\in W^{\mathcal{F}}$ with $\mathcal{A}(\vec{U})=\vec{u}$, so Axiom $IGN$ is sound.

The soundness of $A_1$-$A_5$ has been proven in [9]. The soundness of $B_1$-$B_6$ has been proven in [7]. The soundness of $A_\neg$, $A_\wedge$ and $A_{[]}$ and the deduction rules is obvious. Therefore $\mathsf{L}_{BC}$ is sound with respect to basic causal plausibility models.  $\square$

**Proof of Lemma** 1 Suppose $[\vec{X}=\vec{x}]V=v\in\Gamma$ and $\vec{U}=\vec{u}\in\Gamma$, then by Axiom $B2$, $\neg K\neg(\vec{U}=\vec{u}\wedge[\vec{X}=\vec{x}]V=v)\in\Gamma$. Then by Axiom $SD$, $K^{\vec{U}=\vec{u}}[\vec{X}=\vec{x}]V=v\in\Gamma$, thus $f_V^\Gamma(\vec{U}=\vec{u},\vec{X}=\vec{x})=v$. On the other hand, if $[\vec{X}=\vec{x}]V=v\not\in\Gamma$, then by $A_1$ and $A_2$, there is $v'\in\Sigma$ with $v\neq v'$ such that $[\vec{X}=\vec{x}]V=v'\in\Gamma$. Then by the same steps $K^{\vec{U}=\vec{u}}[\vec{X}=\vec{x}]V=v'\in\Gamma$, therefore $f_V^\Gamma(\vec{U}=\vec{u},\vec{X}=\vec{x})\neq v$.  $\square$

**Proof of Lemma** 3 By Axiom $IGN$, $\neg K\neg(\vec{U}=\vec{u})\in\Theta$, so by Axiom $KB$, $\neg B^{\vec{U}=\vec{u}}\perp\in\Theta$. By Lemma 2, $Min_{\leq^\Gamma}|\vec{U}=\vec{u}|_\Gamma\neq\varnothing$. Therefore there is at least one assignment in $W^\Gamma$ with $\vec{U}=\vec{u}\in\Theta$. Let $\phi$ be any formula in $\Theta$, by Axiom

$B2$, $\neg K \neg (\vec{U}=\vec{u} \wedge \phi) \in \Theta$. Then by Axiom $SD$, $B^{\vec{U}=\vec{u} \wedge \neg \phi} \perp \in \Theta$. By Lemma 2, $Min_{\preceq r} |\vec{U}=\vec{u} \wedge \neg \phi|_r = \varnothing$. Therefore, for any $\Theta'$ with $\vec{U}=\vec{u} \in \Theta'$, $\neg \phi \notin \Theta'$. Since $\Theta'$ is maximal consistent, $\phi \in \Theta'$. Thus $\Theta = \Theta'$.     □

**Proof of Proposition 5(Decidability)** Given a signature $\mathcal{S}=(\mathcal{U}, \mathcal{V}, \Sigma)$. Let $\langle \phi \rangle = \{X \in \mathcal{U} \cup \mathcal{V} \mid X \text{ occurs in } \phi\}$. Let $\mathcal{S}_\phi = (\mathcal{U}_\phi, \mathcal{V}_\phi, \Sigma_\phi)$, where $\mathcal{V}_\phi = \mathcal{V} \cap \langle \phi \rangle$; $\mathcal{U}_\phi = \mathcal{U} \cap \langle \phi \rangle \cup \mathcal{U}^*$ where $\mathcal{U}^*$ is a set of fresh variables with $|\mathcal{U}^*|=|\langle \phi \rangle|$; $\Sigma_\phi$ is a finite subset of $\Sigma$ and contains all the values that appear in $\phi$. Note that $\mathcal{S}_\phi$ is finite.

We construct a finite model $M = \langle \mathcal{S}_\phi, \mathcal{F}_\phi, \leq_\phi, \mathcal{A}' \rangle$ based on $\mathcal{S}_\phi$. First, we define $\mathcal{F}_\phi$: since $\phi$ is satisfiable in $M$, there exists an ordering $\prec$ among the endogenous variables in $\mathcal{V}$ such that if $X \prec Y$, then the value of $F_X$ is independent of the value of $Y$. Let $Pre(X) = \{Y \in \mathcal{U} \mid Y \prec X\}$ and let $D(X) = \{U \in \mathcal{U} \mid U \text{ influences the value of } X\}$. For convenience, we only allow $f_X$ to take the values of variables in $\mathcal{U} \cup Pre(X) \cup D(X)$ as parameters. Then for any endogenous variable $X \in \mathcal{U}$, we define $f'_X$ as follows: Induction on $\prec$. Let $f'_X(\vec{u}, \vec{a}) = f_X(\vec{u}, \vec{b})$ where $\vec{a}$ is the values of variables $\mathcal{V}_\phi \setminus \{X\}$ and $\vec{b}$ is the values of variables in $Pre(X)$, $\vec{u}$ is the values of variables in $\mathcal{U}$. If $X$ is $\prec$-minimal, then $f'_X(\vec{u}, \vec{a}) = f_X(\vec{u})$. Inductive: let $f'_X(\vec{u}, \vec{a}) = f_X(\vec{u}, \vec{b})$ where $\vec{a}$ is the values of variables $\mathcal{V}_\phi \setminus \{X\}$ and $\vec{b}$ is the values of variables in $Pre(X)$. For any $Y \in Pre(X)$, if $Y \in Pre(X) \cap \mathcal{V}_\phi$, then the value of $Y$ in $\vec{b}$ is the value of $Y$ in $\vec{a}$. If $Y \in Pre(X) \setminus \mathcal{V}_\phi$, then the value of $Y$ in $\vec{b}$ is $f'_Y(\vec{u}, \vec{a})$(By I.H, $f'_Y(\vec{u}, \vec{a})$ has been defined).

For exogenous variables: There are two cases: (1) $D(X) \subseteq \langle \phi \rangle$, $f^\phi_X(\vec{u'}) = f'_X(\vec{u})$ where $u'$ denotes the values of exogenous variables in $\mathcal{U}^\phi$ and $\vec{u}$ denotes the value of variables in $\mathcal{U}$, and $\vec{u'}$ agrees $\vec{u}$ on the values of variables in $D(X)$; (2) there is a non-empty exogenous variable set $\vec{B} = D(X) \setminus \langle \phi \rangle$, then we pick a fresh variable $U_X \in \mathcal{U}^*$ and a value $u_x$, let $f^\phi_X(\vec{u'}, u_X) = f'_X(\vec{u})$. Then $U_X = u_X$ iff $\vec{B} = \vec{b}$ where $\vec{b}$ component in $\vec{u}$.

Let $\mathcal{F}_\phi = \{f^\phi_X \mid X \in \langle \phi \rangle\}$, we define $\leq_\phi$ as follow: for any $\mathcal{A}'_1, \mathcal{A}'_2$ in $M_\phi$, $\mathcal{A}'_1 \leq_\phi \mathcal{A}'_2$ iff for any $\mathcal{A}_1, \mathcal{A}_2$ in $M$, if $\mathcal{A}'_1(X) = \mathcal{A}_1(X)$ and $\mathcal{A}'_2(X) = \mathcal{A}_2(X)$ for all $X$ appear in $\phi$, then $\mathcal{A}_1 \leq \mathcal{A}_2$. Then we induction on $\phi$. The cases without epistemic operators are similar in [9]. We consider the case that $\phi = B^\alpha \psi$. If $M \vDash \phi$, then $Min_\leq ||\alpha|| \subseteq ||\psi||$. Suppose that $M_\phi \not\vDash \phi$, then there exists an assignment $\mathcal{A}'_n \in Min_{\leq_\phi} ||\alpha||$ and $\mathcal{A}'_n \not\vDash \psi$. Which means there is an assignment $\mathcal{A}_n$ in $M$ such that $\mathcal{A}_n(X) = \mathcal{A}'_n(X)$ for all $X \in |\phi|$ and $\mathcal{A}_n \in Min_\leq ||\alpha||$. Since all variables in $\psi$ occur in $\phi$, we have $\mathcal{A}_n \not\vDash \psi$, and hence, $M \not\vDash B^\alpha \psi$ which is a contradiction.

The other direction: It is easy to extend a finite model to an infinite model by adding infinitely many irrelevant fresh variables and extending the plausibility ordering.     □

# References

1. Baltag, A., van Ditmarsch, H., Moss, L.: Epistemic logic and information update, Handbook of the Philosophy of Information, pp. 361–456. Elsevier Science Publishers, Amsterdam (2008)

2. Baltag, A., Moss, L.S.: Logics for epistemic programs. Synthese **139**(2), 165–224 (2004)
3. Baltag, A., Smets, S.: Probabilistic dynamic belief revision. Synthese **165**(2), 179–202 (2008)
4. Baltag, A., Smets, S.: A qualitative theory of dynamic interactive belief revision. Logic Found. Game and Decis. Theory (LOFT 7) **3**, 9–58 (2008)
5. Barbero, F., Schulz, K., Velázquez-Quesada, F.R., Xie, K.: Observing interventions: a logic for thinking about experiments. J. Logic Comput. (2022)
6. van Benthem, J.: Logical Dynamics of Information and Interaction. Cambridge University Press, Cambridge (2011)
7. Board, O.: Dynamic interactive epistemology. Games Econom. Behav. **49**(1), 49–80 (2004)
8. Fagin, R., Halpern, J., Moses, Y., Vardi, M.: Reasoning About Knowledge. MIT Press, Cambridge (1996)
9. Halpern, J.Y.: Axiomatizing causal reasoning. J. Artif. Intell. Res. **12**, 317–337 (2000)
10. Halpern, J.Y.: From causal models to counterfactual structures. Rev. Symbol. Logic **6**(2), 305–322 (2013)
11. Ibeling, D., Icard, T.: On open-universe causal reasoning. In: Globerson, A., Silva, R. (eds.) Proceedings of the Thirty-Fifth Conference on Uncertainty in Artificial Intelligence, UAI 2019, Tel Aviv, Israel, 22–25 July 2019. Proceedings of Machine Learning Research, vol. 115, pp. 1233–1243. AUAI Press (2019). https://proceedings.mlr.press/v115/ibeling20a.html
12. Ibeling, D., Icard, T.: Probabilistic reasoning across the causal hierarchy. In: Proceedings of the AAAI Conference on Artificial Intelligence, vol. 34(6), pp. 10170–10177 (2020)
13. Khan, S.M., Lespérance, Y.: Knowing why-on the dynamics of knowledge about actual causes in the situation calculus. In: Proceedings of the 20th International Conference on Autonomous Agents and MultiAgent Systems, pp. 701–709 (2021)
14. Lin, H., Kelly, K.T.: A geo-logical solution to the lottery paradox, with applications to conditional logic. Synthese **186**, 531–575 (2012)
15. Lin, H., Kelly, K.T.: Propositional reasoning that tracks probabilistic reasoning. J. Philos. Log. **41**(6), 957–981 (2012)
16. Meyer, J.J.C., Hoek, W.V.D.: Epistemic Logic for AI and Computer Science. Cambridge Tracts in Theoretical Computer Science, Cambridge University Press (1995)
17. Pearl, J.: Causal diagrams for empirical research. Biometrika **82**(4), 669–688 (1995)
18. Spirtes, P., Glymour, C.N., Scheines, R.: Causation, Prediction, and Search. MIT press, Cambridge (1993)
19. Van Ditmarsch, H., van Der Hoek, W., Kooi, B.: Dynamic Epistemic Logic, vol. 337. Springer Science & Business Media, Berlin (2007)
20. Verma, T., Pearl, J.: Causal networks: semantics and expressiveness. In: Machine Intelligence and Pattern Recognition, vol. 9, pp. 69–76. Elsevier (1990)

# Author Index

Printed in the United States
by Baker & Taylor Publisher Services